LEAD IN THE ENVIRONMENT

LEAD IN THE ENVIRONMENT

Proceedings of a conference held at the Zoological Society of London, Regent's Park, London, N.W.1, organized by the Institute of Petroleum, in association with the Chemical Society and the British Occupational Hygiene Society

Edited by
PETER HEPPLE

APPLIED SCIENCE PUBLISHERS LTD
BARKING, ESSEX

on behalf of

THE INSTITUTE OF PETROLEUM
LONDON

APPLIED SCIENCE PUBLISHERS LTD
RIPPLE ROAD, BARKING, ESSEX, ENGLAND

*The symbol I.P. on this book implies that the
text has been officially accepted as authoritative
by the Institute of Petroleum, Great Britain.*

First published 1971
by The Institute of Petroleum

Reprinted 1973 using new litho plates

ISBN 0 85334 485 X

Printed in Great Britain by Galliard Limited, Great Yarmouth, Norfolk, England

Contents

Opening Address

By Sir ERIC ASHBY, F.R.S.

(*Master of Clare College, Cambridge; Chairman, Royal Commission on Environmental Pollution*)

FIRST of all, it is essential for me to remove any misunderstanding as to why I am here. I am not an expert on lead, and the people who organized this conference know this. I am here, just as the chairman said, to "say a few words", which is an occupational risk of people who reach my stage of senescence and hold jobs of responsibility. The one safeguard in asking people to speak who have left academic work and become administrators is that they are incapable of any sustained intellectual effort, and therefore they can be relied upon not to go on too long. The sole justification for my being here is to associate the Royal Commission on Environmental Pollution with this conference. This I very gladly do, for the Royal Commission is not a body which provides expert opinion; it consists of eight more or less respectable British citizens, unpaid, who act in a quasi-judicial capacity, advising the Government what the priorities are for abating pollution, and acting as a sort of watchdog for the public. A very important part of their work is to interpret to the public the findings of experts. It is for that last reason that I am particularly glad to be invited to open this meeting.

Mark Twain, writing about some scientific work on the Mississippi, said the fascination about science was that you got such great returns of conjecture for such a small investment in fact. This is a good motto for any body which is concerning itself with pollution. Most of the problems of pollution are thought, quite rightly, to be concerned with economics and politics rather than science. But it is true that as soon as you begin to ask perceptive questions you do run up against an under-investment of fact. This is particularly true of the problems of pollution by lead. Of course, the paramount problem, which I suppose we would be unanimous about, is to understand the long-term effects of exposures to low concentrations. This is true of all pollution problems. But I feel, and this is the only substantial point I would like to make in these remarks, that it is enormously important that the problems which you are going to talk about today should be put across to the public. The public are not going to read the excellent preprints which have been circulated today. They are written at what I would call "Third Programme level", which means that people like myself, who might still just get through "A" levels in chemistry and physics, can read them. But they are not going to be read by the kind of people who subscribe to the *Daily Mail*. These are the people who, in the end, determine political decisions. For although politicians are anxious to take the advice of their experts in the end public opinion exerts its pressure upon them, and it is the politicians' job to respond to public opinion. So there is a very urgent need to put out (and this is what the Royal Commission tries to get bodies like yourselves to do) statements about pollution which the public can easily understand and which are authoritative. They should state first of all the indisputable facts, and then record quite frankly the things we still need to know before there could be any certainty in decision making.

The indisputable facts begin by our acknowledgment that lead is a poisonous substance; that it exists in the air, the water, and the soil; that people take in 300–400 micrograms a day and absorb 10 per cent of it; and that it is increasing in concentration in the air. In a paper just published in the *Journal of the Royal Institute of Chemistry* there is an interesting graph showing that the use of tetra-ethyl lead has gone up some 400 per cent since the end of the last war, and the number of miles per car travelled with lead additives has gone up 83 per cent. Another important fact, which I hope will be discussed at this meeting, is the hint of long-term effects of low concentrations evident from the work that has been done on the effect of lead on the activity of ALA dehydrogenase. These are the kinds of facts which ought to be not only discussed among experts but interpreted to the public.

At this stage your real problem begins. For the scientist hates to reach decisions until he has got enough data; the politician has to reach a decision before he has enough data. Therefore, a risk has to be taken. I think we ought, therefore, to put frankly in front of the public the way in which the risks of these pollutants are calculated, to explain the voluntary co-operation between government and industry which has been successful in dealing with some kinds of pollution in this country, and the kinds of legislation and further voluntary co-operation which should be sought. This need to take a risk is nothing new. The most obvious example of it is the way in which we as a nation are prepared (although heaven knows why) to design and market cars which can run at a speed which kills 7000 people a year and injures 350,000. That is the price we are prepared to pay for the luxury of travelling at that speed in that kind of instrument. Yet we know perfectly well that no government which tried to introduce an upper speed limit of, say, 25 mph for all vehicles, everywhere, at all times (which after all is no more onerous than the travel conditions which faced Dr Johnson when he went to Scotland from London)—no government prepared to do that would last for five minutes. So we are prepared to take risks, and the risk that we need to take in most problems of pollution are a great deal smaller than the risks we take on our roads. This is the kind of cold statement that ought to be put in front of the public by experts like yourselves in language that can be understood. They will then realize that unless lead is outlawed altogether (and it is a very valuable metal) it has to be used with a certain minimal risk. The risk can be calculated and people ought to know what it is. It will turn out, compared with the example I gave you, to be pretty small.

I hope, therefore, that there can be some serious discussion (perhaps, Mr Chairman, by the Institute itself) of a way to inform not only the kinds of people gathered here this morning about the issues that you will discuss but the public themselves. This would counteract some of the hysterical, and sometimes downright mischievous, statements made by people who put together little fragments of fact and draw totally unwarranted conclusions from them. I would like to end with one other, more technical, plea. Is it possible for those who work on lead pollution to try to simplify the units used to express concentrations of lead? Micrograms per millilitre, micrograms per cubic metre, parts per million, grams of lead per litre, grams of lead per American gallon, sometimes even grams per mile travelled by a car. This is very confusing, even to people whose science is at "A" level standard, like myself. With a little more care the jargon used by scientists could be made easier to follow, and this would help journalists, who do work very hard trying to interpret these matters. It would make easier their job of putting clearly the problems of lead pollution in front of the public.

So, thank you very much, Mr Chairman, for giving a representative from the Royal Commission a chance to meet you and to assure this audience that we are doing our best to interpret to the public the state of the British environment. If you do not read much about our activities it is because we are all the time in quiet and amicable contact with government departments. One of the things we suspect is that as soon as a government department knows that the Royal Commission has become interested in a pollution problem they themselves begin to consider possible measures of control! And this is as it should be. We work in this curious pragmatic British way, which is not very ostentatious. And as part of our activities, I do now ask a body like yourselves to play a part, not only in stimulating investment in fact (which Mark Twain quite rightly said is too small) but at the same time in informing the public what is known, what is not known, where one can be sure, where one can only guess. I now have much pleasure in handing you over to your proper Chairman.

Sources of Lead in the Environment

By R. L. STUBBS, O.B.E.

(Director-General, Lead and Zinc Development Associations)

LEAD is a widespread constituent of the earth's crust. It has always been present in soils; in rivers, lakes, and seas; in the air, following the burning of wood and coal; and in plants, both edible and inedible. Lead in soil ranges from 2 to 200 ppm and averages 16 ppm. However, lead compounds are generally insoluble and so poorly transferred from one medium to another. As a result, the concentration in natural bodies of water is extremely low—between 0·001 and 0·01 ppm.

Throughout history, lead has been well known and widely used by mankind and so over a long period has been dispersed by man into the environment. Today lead is a major material of industry, far more widely used than is perhaps generally realized. However, it is of particular significance to this conference that lead is less readily released into the environment than most other metals in everyday use, since in many of its applications it is virtually indestructible.

Before reviewing the processing and current uses of lead and how it may enter the environment, a few facts about world consumption and the situation in the U.K. need stating.

The current world consumption of refined lead is about 4 million tonnes/year and has grown during the past decade at an average rate of $3\frac{1}{2}$ per cent/year. In 1970 the U.S.A., the largest user, took about 1·13 million tonnes, the U.S.S.R. is thought to have consumed about 500,000 tonnes, Germany about 310,000 tonnes, and the U.K. about 262,000 tonnes. Consumption in the U.K. has fallen in recent years.

Deposits of lead ores are widely distributed throughout the world and important mines are still operated in many of the oldest industrial countries. British mine output was the largest in the world until about 100 years ago, but today is very small. In recent years new ore bodies have been opened up in many countries. In the last year or two the U.S.A. moved ahead of Australia and Canada to become, once again, the largest mine producer in the world and, following new discoveries, Southern Ireland is now an important source of lead ores.

However, an important feature of lead is the widespread recovery of scrap, sometimes referred to as the overground mine. For the world as a whole about a quarter of all lead material for refining is scrap. In the U.S.A. and U.K. about half the refined lead production is derived from scrap—a good example of recycling which many ecologists overlook.

Most industrialized countries have a substantial production of refined lead derived from ores and scrap and this is supplemented by the re-use of other scrap which is simply re-melted, sometimes after adjustment of its alloy content.

I

In the U.K. the total consumption of lead in all forms amounted to about 350,000 tonnes in 1970 and this comprised about 262,000 tonnes of refined lead and about 88,000 tonnes of scrap and re-melted lead used direct by consumers. While half of the U.K. production of refined lead comes from scrap, the other half comes from imported bullion, but the U.K. both imports and exports lead, as can be seen in Appendix 1, which summarizes the supply situation for the U.K.

Release of Lead into the Environment

There are various ways in which lead can enter the environment: from primary and secondary smelting operations; from fabricating processes; as a result of ways in which it is used; in the combustion of coal and other natural products; and in the disposal of discarded or unwanted materials incorporating lead products. The entry of lead into the environment from production and manufacturing processes tends to be concentrated in a relatively few works governed by strict regulations which set safe levels for emissions and lay down safe working procedures for handling lead materials. On the other hand, when lead is used in such a way that it is eventually released into the atmosphere, release is gradual and over a relatively wide area, so that concentrations remain small. This is even more so when worthless old materials containing a little lead are destroyed by burning. Lead has been discharged into the atmosphere as long as coal has been burnt, and it is said that each cigarette smoked releases about one microgram of lead.

Production and Fabrication of Lead

Lead smelting was once the main source of lead released into the atmosphere, but smelting processes and the control of fume emission have been greatly improved during the last 50 years. It is now possible to collect virtually all the fume and dust and to recover its lead content. As a result, the emission of lead into the atmosphere from U.K. lead refineries is now very small and closely monitored. There are about a dozen lead refineries in Britain spread fairly widely throughout the country. Some are comparatively small.

In lead fabricating, *i.e.* rolling and extrusion and the manufacture and casting of lead alloys and the manufacture of most lead compounds, the operations are carried out at low temperatures with little risk of fume formation. The few exceptions are referred to later on.

Uses of Lead

The uses of lead are diverse and numerous and fall into two groups—those in metallic form and those in the form of chemical compounds. Details of U.K. lead consumption are given in Appendix 2.

In Britain the metallic uses of lead take two-thirds of the total consumption, the most important being for batteries, cable sheathing, sheet and pipe, alloys of various types, lead shot, and a large number of minor miscellaneous applications. Most are based on the high corrosion resistance of the metal, and only a minute proportion of lead in these forms is ever released into the environment. The other third of the lead consumed in Britain is in the form of lead compounds. Most important are the oxides for lead batteries, followed by lead additives for gasoline, lead pigments of various types, and numerous lead chemicals used in the manufacture of plastics, glasses, glazes and other products. In many of these uses lead compounds are either isolated, *e.g.* the oxides in batteries, or chemically combined into a very insoluble form, *e.g.* in glasses and glazes. Some uses of compounds result in lead being released into the environment.

With the exception of the U.S.A., the pattern of lead usage is much the same in the main industrial countries, and the metallic uses have tended to increase more than those of most lead compounds, except lead additives for gasoline. In the U.S.A. batteries and lead additives are by far the major uses.

Batteries. The largest and fastest growing outlet for lead throughout the world is in lead batteries—for motor cars, electric vehicles, trucks and other appliances, for emergency lighting, and so on. About 40 per cent of world consumption of refined lead goes into batteries, and in Britain they take about 95,000 tonnes/year. About half the lead used in batteries is in the form of lead alloys and the other half in the form of oxides. Batteries are, in fact, enclosed packages of lead and none of the lead in them can enter the environment during their use. Virtually all the lead used in batteries is recovered, largely within the short period of three to four years. Battery scrap goes back eventually to secondary smelters and refiners, who have special equipment for handling the large arisings of this material. Neither battery manufacture nor treatment of scrap batteries need cause any pollution problems.

Cable sheathing. In Britain, cable sheathing was for a long time the largest application of lead and it is only in the last few years that it has been overtaken by batteries. Cable sheathing now takes about 60,000 tonnes/year. There must be thousands of miles of lead-sheathed cable in use today, most of it buried in the ground, and the lead is usually covered by an outer wrapping. Lead sheathing is virtually indestructible and the chances of it entering the environment are negligible.

Sheet and pipe. Another long-standing traditional use of lead is for sheet and pipe, which nowadays takes about 53,000 tonnes/year in Britain. They are used mainly in building and, to a lesser extent, also in chemical plants. The demand for lead pipe for plumbing is small compared with what it was in the past, since it has been largely superseded by cheaper alternatives. Much old lead pipe has been recovered as scrap in urban redevelopment and slum clearance but a great deal is still in use, both for waste and water pipe. Lead water pipes can only cause problems when the water carried in them is plumbo-solvent. Through long experience, water authorities throughout the country know when the use of lead pipe cannot be permitted. Lead sheet, used in building for flashings, roof coverings, and in various other ways, has a very high resistance to atmospheric corrosion. The patina that forms on lead on exposure to the atmosphere is strongly adherent and highly insoluble and any loss on weathering, even after 100 years or more, is negligible. The main use of lead sheet and pipe in chemical plant is for handling sulphuric acid. There is a very slow loss of lead into the acid, largely due to erosion; very little chemical solution takes place.

Other metallic uses. There are many other uses of lead in metallic form, ranging from lead shot and fishing weights to yacht keels, but the most important in tonnage are lead alloys such as printing metals and solders. There is a large mass of printing metal in constant use, but it moves within a closed circuit between refiners and printing works and these is little risk of any loss into the environment. Lead-containing solders have many applications, mostly in factory processes. There would seem to be no problems in the use of solders in the electrical industry or in plumbing. The amount of solder which comes in contact with the contents of food and beverage cans is very small and experience has shown that virtually no lead is picked up by the contents, even when the inside of the can is unlacquered.

Many alloys contain small quantities of lead, such as leaded brasses and leaded steels, in which the total quantities of lead used are small. Some of this lead might enter the environment when the basic material is being re-melted at high temperatures in scrap recovery plants but most brass founders are aware of the danger and exercise the necessary fume control.

Pigments. White lead—basic lead carbonate—was, in the past, extensively used as a pigment for paints and was a major outlet for lead, taking up to 30,000 tons/year in the 1930s. Production has been declining sharply since World War II and now takes only 3000 tons/year. Only part is used in paints, mainly in wood primers, the other major use being as a stabilizer in PVC sheathing for cables.

The traditional white lead paint weathered by "chalking" and so some of the lead carbonate would have been washed away by rain, to be dispersed into the environment, but in their modern use as priming coats for woodwork, they are, of course, sealed off by other paints. All paints containing lead must be clearly marked; their use for painting toys is now banned, and indeed there have been voluntary restrictions in Britain for a very long time. Nevertheless, the widespread use of white lead paints in the past, both indoors and outdoors, means that there are many old buildings in which these paints are still to be found, even though they may have been covered with other paints. This may not be regarded as an important source of lead in the environment in Britain, but in the U.S.A., where old timber buildings are much more common, lead paints constitute a recognized hazard because of flaking and chalking, and from the lead-containing particles released when old painted timber is burnt.

Red lead and the more recently developed calcium plumbate are widely used as rust-inhibiting pigments in primers for iron and steel, where they are usually sealed off by undercoats and top coats. Another important pigment is yellow lead chromate, which is used in exterior paint and, to a lesser extent, in paints and plastic compounds for road marking. However, lead chromate is among the least soluble of all lead compounds. These lead-pigmented paints, like other paints, tend to be eroded, and therefore in a small way can be said to constitute a source of lead entering the environment.

Lead in glazes, glasses, plastics, etc. Lead bisilicate is widely used in glazing ceramics. There is no risk of any harmful release of lead from these glazes when they are properly formulated and fired, since rigid quality control is exercised by manufacturers of pottery and the lead is chemically combined in a very insoluble form. Most countries have national standards which impose tight limits to ensure that decorative glazes containing metals, including lead, are safe to use. Consumption of lead for glazes in Britain is about 3000 tonnes/year, and another 3500 tonnes of lead compound go into the manufacture of glass, not only of crystal glass, but also of fluorescent lights and television tubes. As in glazes, the lead in glass cannot be released accidentally.

Lead compounds, principally tri-basic lead sulphate, are added as stabilizers to PVC plastics used for making household objects, water pipes, gutters, and downpipes, etc. Extensive tests carried out by water and other authorities have demonstrated that they are safe in service with all normal potable waters.

There are many minor uses of lead compounds—taking possibly 3000 tonnes of lead/year—such as driers for paints and printing inks, in pyrotechnics, fungicides, linoleum, anti-fouling paints, and insecticides. Powdered metallic lead is also used as a pigment for priming paints for iron and steel. Small quantities can enter the environment from these uses, but they are well scattered and unlikely to cause significant lead concentrations.

Lead additives in gasoline. Tetraethyl and tetramethyl lead are added to gasoline because this is the most convenient and economic method of increasing the octane ratings of all grades of gasolines; in particular, they are usually responsible for obtaining the final six to eight octane numbers in the high-grade petrols on which the development of engines of high compression ratio has been based.

The U.K. is the largest manufacturer of lead additives in Europe. About 40,000 tonnes of lead/year are being used to make these additives and about 80 per cent of the total U.K. production is exported. Of the 9000 tonnes of lead at present being added to petrol in the U.K., about a quarter is retained in the motor car itself—in the exhaust system, the engine oil, and filters. The rest of the lead is discharged in the exhaust with the gaseous products of combustion, the lead itself being mainly in the form of fine particles of lead compounds, which constitute about one-third by weight of the solids in the exhaust. Half the lead-containing particulate matter falls to the ground within a few hundred feet of roadways and is then washed away and dispersed in the soil and drains. Finer particles are dispersed in the atmosphere and may be carried considerable distances by air movements before they are eventually deposited.

While the total consumption of lead additives throughout the world has been increasing year by year, it should not be assumed that the concentrations of lead in air in urban areas are increasing in parallel. Most available information suggests that the concentrations are rising only slowly, if at all, *i.e.* most of the suspended lead particles are dispersed as fast as they are released. Although this meeting is concerned with lead, it should perhaps be stressed that, while there is naturally concern about the concentrations of lead in the atmosphere that arise from this use of lead, the major pollutants in automobile emissions are carbon monoxide, unburnt and partly burnt hydrocarbons, and oxides of nitrogen.

Lead additives are used in all industrial countries, but the amounts vary widely. Though U.K. petrol consumption rises all the time, the total weight of lead added in 1970 was no more than five years earlier, presumably because higher octane petrols are being produced and needing less lead additive to bring them up to the required final grade. In the U.S.A., on the other hand, lead consumption for this purpose has continued to grow and in 1970 amounted to about 250,000 tonnes. World total consumption of lead for additives has been estimated at 350,000 tonnes in 1970.

Conclusion

This outline of the use of lead shows its importance as an industrial material and the considerable contribution it makes to modern living through a diversity of uses. Like many other materials in everyday life, there are times in its production, processing, and use when care must be taken to avoid it polluting the environment and presenting a hazard to public health. It has been well known for a very long time that lead is toxic in some of its forms, but so are many other commonplace materials and certainly more is known about the effects of exposure to lead than of most other materials.

APPENDIX 1

U.K. Lead Supplies

	Tonnes		
	1960	1965	1970
U.K. refined lead from scrap	92,000	127,000	147,000
U.K. refined lead from imported bullion	51,000	45,000	140,000
Imported refined lead	160,000	178,000	97,000
Total refined lead supplies	303,000	340,000	384,000
Exports of refined lead and alloys	16,000	31,000	112,000
Net refined lead supplies for U.K. consumption	287,000	309,000	272,000
Consumption of scrap and re-melted lead	98,000	123,000	88,000
Total U.K. lead consumption	385,000	432,000	360,000

APPENDIX 2

U.K. Uses of Lead

	Tonnes		
	1960	1965	1970
Metallic forms			
Cables	97,000	135,000	60,000
Batteries	40,000	43,000	49,000
Shot	6,000	6,000	6,000
Sheet and pipe	75,000	70,000	53,000
Foil —	4,000	1,300	400
Collapsible tubes —		2,400	2,000
Other rolled and extruded products	9,000	6,000	2,000
Solder	15,000	16,000	14,000
Alloys	22,000	24,000	20,000
Miscellaneous	17,000	21,000	22,000
Chemical forms			
Battery oxides	37,000	43,000	46,000
Lead alkyl anti-knock compounds	27,000	37,000	40,000
White lead	8,000	4,000	3,000
Other oxides and compounds	27,000	27,000	32,000

Discussion

Dr J. R. Glover (Welsh National School of Medicine): When you say that 9000 tons of lead are added to petrol in the U.K. each year, what about the quantity of the lead that is imported each year in leaded petrol?

Professor D. Bryce-Smith (University of Reading): Is petrol containing lead imported into the U.K.?

G. S. Parkinson (Shell-Mex and B.P. Ltd) (written answer): In 1970, just over 14 million tons of motor spirit was sold in the U.K., of which 11·2 million tons was home-refined. In the same year 7500 tons of lead was used in motor spirit manufactured in the U.K. and a small proportion of this was exported. The quoted figure of approximately 9000 tons for total lead in all U.K. gasoline is therefore not unreasonable.

Airborne Lead and its Uptake by Inhalation

By P. J. LAWTHER,* B. T. COMMINS, J. McK. ELLISON,
and B. BILES

(*MRC Air Pollution Unit, St Bartholomew's Hospital Medical College,
Charterhouse Square, London EC1M 6BQ*)

INTRODUCTION

LEAD has long been recognized as a poison and much work has been done on its distribution and toxicology. During the past century the industrial causes of classical gross plumbism have been removed but concomitantly the diagnosis of lead intoxication has become better and the detection of lead more sensitive; with these advances the possible effects of long exposure to low concentrations has been considered. It is only in the last few years, however, that there has been any serious suggestion that with the increasing quantities of leaded petrol being used the concentration of lead in the general atmosphere of large towns might have risen to the point that it constituted a public health hazard. Hitherto there has been little information available on which to assess the danger: the purpose of this paper is to present measurements and observations which it is hoped will help to fill the gap, and to review these and other data that have a bearing on whether or not a real public health problem exists, and whether exposure to airborne lead is increasing.

The data are of three different types: (1) measurements of airborne concentrations of inorganic lead compounds, in a busy street and at a "control" site away from traffic, in London during the year 1962–63[1] and again in 1971, and also some daily measurements at the control site since 1965; (2) measurements of the blood lead concentrations in a group of workers particularly exposed to pollution by road vehicles, namely, London taxi drivers, using day workers as a "test" group, and night workers, who are exposed at a time when there is less traffic, as "controls"; and (3) observations on the visual appearance, under the electron microscope, of particles collected: (a) from the general atmosphere of London and from a busy street there; and (b) from the exhaust gases of petrol and diesel engines. It will be evident from this list that we have measured only particulate, inorganic lead. It seems to be agreed that most of the lead in air is in this form, although in the air of Los Angeles it is claimed that between 2 and 10 per cent may be present as lead alkyl vapour[2] and Laveskog[3] has suggested even higher figures for Stockholm. Toxicologically, however, the lead alkyls differ from inorganic lead compounds and it is therefore not unreasonable to keep them separate here.

* Director, MRC Air Pollution Unit.

8

AIRBORNE LEAD IN LONDON

Sites: Measurements were made at two sites, one in Fleet Street and one in the sampling house on the roof of our laboratories. The street site was the same as had been used in a previous survey in 1962–63[1], which made it possible to compare measurements made then and in 1971. The roof site is about 70 m from the street, measured horizontally, and about 20 m above it; it was chosen because it is well away from streets and because we have made other measurements there over the past six years so that it was again possible to compare results over a period. At both sites total particulate matter, particulate lead, the polycyclic aromatic hydrocarbon coronene, carbon monoxide, and nitric oxide were measured. Coronene was chosen because it is emitted mainly by motor vehicles and is stable, easily measured, and efficiently collected by filtration.[4]

Techniques of sampling: Since the airborne concentration of the pollutants is higher in the street than on the laboratory roof, different sampling methods were used. In Fleet Street particulate matter was collected by drawing about 15 litres of air per minute through weighed glass fibre filters, which were mounted in holders made for the purpose: the exposed area was a circle of diameter 10 cm, facing downwards. Each day, Monday to Friday, the sampling pumps were started by time switch at 08.00 hours and stopped at 19.00 hours, and the soiled filter was then weighed. To simplify handling, the filters were mounted between an annulus of paper on the upstream side and a backing filter on the downstream side.[5]

The samples from the roof of the laboratory were collected by high volume sampler (flow-rate approx $1 \cdot 5$ m³/min) on to weighed glass fibre sheets of exposed area 20 cm $\times$ 25 cm, on the same time schedule as at Fleet Street. Although clearly there are objections to using sampling methods that are not strictly the same at the two sites, the face velocity is in both instances considerably higher than the falling speed of any particles that will remain airborne, so that the samples collected should be closely comparable.

Two other types of sample were collected in the roof sampling house, both of particulate matter collected on weighed glass fibre filters. For one, the exposed area was a circle of diameter 10 cm and air was drawn through this at 3 l/min for one to three weeks; for the other, the exposed area was a circle of 5 cm diameter, the sampling rate was $0 \cdot 1$ m³/min, and four 24-hr samples were collected each week, starting at 12.00 hours on Monday and ending at 12.00 hours on Friday.

Lead was also estimated in old samples that had been obtained for other purposes over the years 1965–71 in the roof sampling room and had been stored. These had been collected from 09.00 hours to 10.00 hours, by high volume sampler as described above (flow rate $1 \cdot 5$ m³/min, area of collector 20 cm $\times$ 25 cm); this series of measurements continued into the period of the present study and is used in Table I below giving the 1971 data, as well as in Table III.

Analysis

Lead: Total or partial samples were extracted with nitric acid and lead estimated colorimetrically with dithizone using the procedure described in our earlier Fleet Street study.[1]

TABLE I

Measured Concentrations of Pollutants in Fleet Street and on Medical College Roof 1971

	Fleet Street 8 a.m.–7 p.m. weekdays					Medical College 8 a.m.–7 p.m. weekdays					Medical College Pb μg/m³		
Date	Pb μg/m³	Smoke μg/m³	CO ppm	NO ppm	Coronene μg/100 m³	Pb μg/m³	Smoke μg/m³	CO ppm	NO ppm	Coronene μg/100 m³	24-hr weekdays	24-hr continuous	9–10 a.m. weekdays
April 30/3–30/4/71	4·9*	441*	—	—	—	0·9*	204*	—	—	—	—	—	1·9
May 3–28/5/71	6·2	468	—	—	—	0·9	184	—	—	—	0·8	—	1·3
June 31/5–2/7/71	5·4	285	17†	0·50†	1·5†	1·0	124	<1·0†	0·03†	0·1‡	0·7	0·6	1·2
July 5–23/7/71	4·9	264	15	0·42	1·5	1·0	154	<1·0	0·03	0·1	0·7	0·8	1·7
August 26/7–30/8/71	5·6	280	15	0·69	1·5	0·9	121	<1·5	0·03	0·1	0·8	0·8	1·7
September 31/8–1/10/71	8·7	366	15	0·81	1·5	1·7	194	<1·5	0·04	0·2	1·3	1·4	2·5
October 4–29/10/71	8·6	323	15	0·71	2·1	2·3	180	<2·0	0·07	0·3	1·1	1·2	3·8
Means§	6·3	347	15	0·63	1·6	1·2	166	<1·5	0·04	0·2	0·9	1·0	2·0
	(7)	(7)	(5)	(5)	(5)	(7)	(7)	(5)	(5)	(5)	(6)	(4)	(7)

* 9 a.m.–5 p.m. or 8 a.m.–5 p.m.
† Incomplete period.
‡ 24/5–28/6/71.
§ Number of months in brackets

Total particulate matter was estimated by weighing samples collected by high volume sampler.

Carbon monoxide was measured by non-dispersive infra-red gas analyser and the results recorded continuously.

Nitric oxide was estimated by Saltzman's method,[6] nitrogen dioxide having first been removed by absorption and the nitric oxide having then been oxidized by bubbling through acid potassium permanganate solution.

To estimate *coronene* the sample was first extracted with hot cyclohexane and separated chromatographically on an alumina column: coronene was then determined by ultra-violet spectrophotometry.[7, 8]

Results

The results of the principal measurements are shown in Table I. In each instance the monthly figures for June, July, and August are very similar but in September and October the concentrations of most pollutants are higher. The table shows that concentrations of pollutants are consistently higher in Fleet Street than at the Medical College, but whereas pollution by smoke in Fleet Street is on average little more than double that in our roof sampling house, the other pollutants, which are particularly associated with road vehicles, are present in much higher concentrations in the street. There are, however, interesting differences among them. If the lead, nitric oxide, carbon monoxide, and coronene were produced exclusively by motor vehicles and if the difference in concentration between street and roof could be attributed only to dilution with air free of these pollutants, then the ratio of the concentrations at the two sites, R,

TABLE II
Apparent Dilution R of Pollutants between Fleet Street and Roof Sampling House

	$R=$(concentration of pollutant in Fleet Street)÷(concentration of pollutant in roof sampling house)				Suggested upper limit of contribution of lead from traffic to lead at roof level $=R(Pb)/R(\text{other pollutant})$*		
Month	$R(Pb)$	$R(NO)$	$R(CO)$	$R(Cor)$	$\dfrac{R(Pb)}{R(NO)}$	$\dfrac{R(Pb)}{R(CO)}$	$\dfrac{R(Pb)}{R(Cor)}$
June	5	17	>17	15	1/3	1/3	1/3
July	5	14	>15	15	1/3	1/3	1/3
August	6	23	>10	15	1/4	2/3	2/5
September	5	20	>10	8	1/4	1/2	2/3
October	4	10	>7	7	2/5	1/2	1/2

* For explanation see text.

would simply be the dilution factor for the air and would be the same for all four. As Table II shows, this is far from being the case: although there are differences between them, NO, CO, and coronene all show consistently higher values of R than does lead. It is, of course, possible that the difference is due to the decomposition of the other pollutants or to their being absorbed by buildings or other surfaces, but all three are relatively stable chemically and this explanation is not very plausible: it would, moreover, imply that the true dilution between street and roof was as low as 4–6, which to us seems unlikely. The more probable explanation is the obvious one: that there are sources of lead, just as there are sources of smoke, other than vehicles, and that these other sources are relatively more important for lead than they are for NO, CO, and coronene. If we assume that none of the NO, CO, and coronene is absorbed or destroyed, then the value of R for lead, R(Pb), divided by the largest value of R for another compound over the same period will constitute an upper limit to the fractional contribution that vehicles make to lead pollution. Values for the quotient R(Pb)/R(NO), etc., are shown in the right-hand columns of Table II. The true value of this fractional contribution is likely to be lower than that obtained in this way, since NO, CO, and coronene come from sources other than internal combustion engines; in particular, in winter they are produced by the burning of coal and oil, which increases in cold weather.

The only series of measurements of airborne lead that we have over an uninterrupted period of years are of samples collected at our laboratory site from 09.00 hours to 10.00 hours, Monday to Friday, since 1965: these are summarized in Table III. From this table it would appear that the concentration of lead in the air had fallen during the past five years. It is, however, well known that accumulation and dispersion of pollutants depend critically on the weather and little can be inferred from observations over a period as short as five years. Nor is it possible to say how typical these figures are of other sites, for our laboratory is in a busy commercial area, criss-crossed with roads, some of considerable commercial importance, and only 70 m from a busy through route taking up to 1200 vehicles/hr. The period 09.00 hours–10.00 hours is one of high emission and poor dispersion, and therefore of high general pollution, and the mean concentrations of lead for this hour are substantially higher than those from 08.00 hours to 19.00 hours, which are in turn rather higher than those for the whole 24 hours.

TABLE III
Concentrations of Lead in Air at Medical College

Mean concentrations of lead (μg/m^3): 9–10 a.m. weekdays at Medical College

Period	1965	1966	1967	1968	1969	1970	1971
Jan.–Mar.	—	2·5	2·5	2·6	1·8	2·5	2·6
April–June	—	2·0	2·1	2·3	1·1	2·0	1·5
July–Sept.	—	2·3	3·1	2·5	2·3	1·0	2·0
Oct.–Dec.	2·9	4·4	3·7	3·5	3·9	2·9	—
Year	—	2·8	2·8	2·7	2·2	2·1	—

During 1962–63 we made measurements of a number of pollutants, including lead, at the same site in Fleet Street,[1] and Table IV shows some of the figures we obtained then alongside figures, as nearly as possible comparable, for 1971. In the earlier period the samples were collected on a quarterly basis and it would therefore not have been possible to make use of all our 1971 data if we had confined ourselves to periods that were strictly comparable: we have therefore used a rather longer period in 1962 than in 1971. The figures show that in 1971 there has been rather less smoke but substantially more lead than in 1962. Although no definite conclusions can be drawn from such limited data, they suggest an

TABLE IV

Comparison of Lead and Smoke Concentrations in Fleet Street in 1962 and 1971

Pollutant	1962 (6 Mar.–31 Aug.)	1971 (30 Mar.–31 Aug.)
Particulate matter (smoke) $\mu g/m^3$	420	352
Particulate lead $\mu g/m^3$	3·2	5·4

increase in lead concentration in Fleet Street over the past nine years; the figures for September and October are considerably higher still and thus point in the same direction. Only time will show whether these higher figures represent a genuine trend or merely a peak in a fluctuating time series, but if the rise is proved true, it will be interesting because for many years Fleet Street has been virtually saturated with traffic during the relevant hours and the number of vehicles involved is unlikely to have risen substantially. According to a recent report,[9] the lead content of petrol is no higher than it was nine years ago, but there have been changes in traffic regulations near the sampling site, and also in vehicles and in the way that they are used. If the increase were due to unusual meteorological conditions or to the use of more fuel, the concentrations of carbon monoxide might have been expected to rise *pari passu* with those of lead, but on the limited evidence available this does not appear to be so. Measurements that we have made in road tunnels in 1959[10] and again (under rather different conditions) recently (unpublished) suggest, however, that the ratio of lead to carbon monoxide in exhaust gases may have increased. For the present it is not possible to resolve the effects of these various factors, and the reason for the difference between the two years remains uncertain.

BLOOD LEAD CONCENTRATIONS IN LONDON TAXI-DRIVERS

This study, which was done in conjunction with Dr R. D. Jones (then a student at the London School of Hygiene), is to be described in detail elsewhere[11] and is only outlined here.

From the figures obtained it was possible to compare the concentration of lead in the blood of taxi-drivers who worked in daytime, when traffic is heavier and when the air may be presumed to be more heavily polluted with fumes from traffic, with that of those who drove at night. As an index of exposure to traffic fumes, carboxyhaemoglobin was also measured in each case. The sample comprised 50 taxi-drivers (aged 22–66), 28 of whom worked during the day and 22 during the night. There is little interchange between these two groups of workers, so that the blood lead figures reflect the effects of long-term exposure to working during the time of day stated.

Analysis of the results showed that, for the smokers, the effects of carbon monoxide in tobacco smoke mask those due to pollution by traffic, but among the non-smokers the concentrations of carboxyhaemoglobin in the day workers reflect the higher concentrations of pollution that are to be expected during the day. Despite this, no appreciable difference between the blood lead concentrations of day drivers and night drivers, whether they smoked or not, was found. From this it would appear that the contribution of inhaled particles to the total blood lead must be small.

ELECTRON MICROSCOPE STUDIES

Sources of samples: Samples were collected (*a*) about 30 cm from the exhaust of a 220-cc diesel engine (running speed 3000 rev/min, about 55 per cent continuous rated load); (*b*) from the air in a small "lock-up" garage into which the exhaust from a 1300-cc petrol engine, burning leaded petrol, was being released (running speed about 2000 rev/min, without external load); and (*c*) from the ambient air in Fleet Street (one example, on membrane filter, at approx 11.30 hours on 21 October 1971 and the other, by thermal precipitator, from approx 11.30 hours to approx 16.00 hours on 8 November 1971).

Methods of collecting samples: Samples were collected by two different methods; (*a*) by thermal precipitator using plugs that had been countersunk to take electron microscope grids, which supported carbon films on which the sample was collected; and (*b*) by drawing air through Sartorius membrane filters (No. 11304). Particles collected by thermal precipitator were observed directly on the carbon films on which they had been deposited; those on membrane filters were coated with a carbon film and the filter then dissolved away, leaving the "extractive replica" on a supporting grid.

Electron microscope technique: By courtesy of the National Physical Laboratory, one sample collected by membrane filter in Fleet Street was examined for lead-containing particles by microprobe analysis, using the AEI EMMA 4. All the electron micrographs shown were taken by direct transmission microscopy using our AEI EM6G: some dense areas in the aggregates showed electron diffraction patterns and so showed themselves to be crystalline, but we have not interpreted these and they are not reproduced here.

Results

Electron micrographs: Figs 1–3 show typical fields under a magnification 25,000, Fig 1 being of exhaust from a diesel engine, Fig 2 that from a petrol engine, and Fig 3 being the mixed pollution found in Fleet Street. At this magnification it is not possible to resolve any detail but the conspicuous feature is that they are so similar in general appearance. In particular, in all these samples the vast bulk of the material is in the form of aggregates and in all three the ultimate particles that make up these aggregates are mainly smaller than 0·1 μm. Figs 4–7 show smaller fields of samples from the same three sources but at higher magnification: in each instance the most conspicuous feature of the picture is again the presence of aggregates of large numbers of smaller particles, which with the higher resolution available can be seen to be rounded and in many instances to look like spheres that have been sintered together. There are, however, differences in detail. The particles that make up the aggregates in the diesel smoke (Fig 4) have an almost diaphanous appearance, which is lacking in those from petrol smoke (Fig 5), although it is difficult to demonstrate this difference on photographic prints. A few of the aggregates in petrol smoke are built up of larger and the more solid-looking spheres (Fig 6) which are conspicuously denser in the middle than at the edges; this is what would be expected of particles made up of material that absorbed homogeneously. The most interesting difference is, however, the presence in the smoke from petrol (both in Fig 5 and Fig 6) of small areas of material that is highly dense to the electron beam. These

FIG 1. Particles from the diluted exhaust of a 220-cc experimental diesel engine.
Magnification × 25,000.

FIG 2. Particles from the diluted exhaust of a 1300-cc car engine using leaded petrol. Magnification × 25,000.

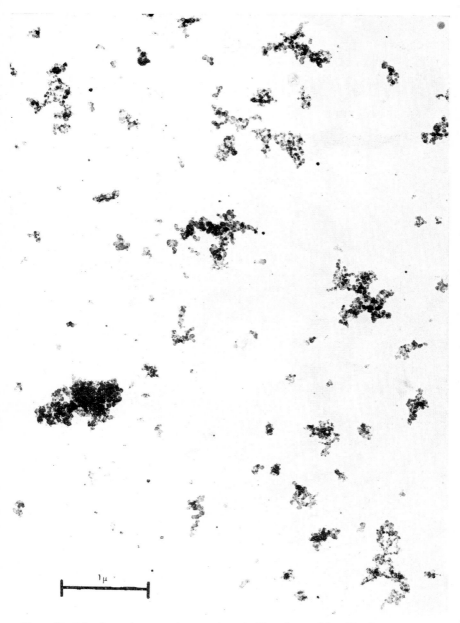

FIG 3. Particles from the general atmosphere in Fleet Street. Magnification × 25,000.

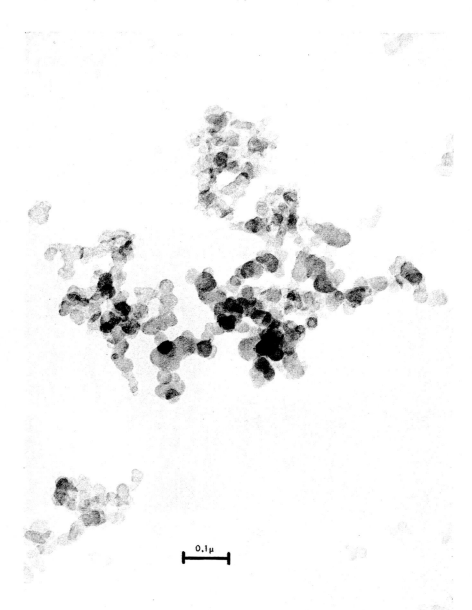

0.1 μ

FIG 4. Particles from the diluted exhaust of a 220-cc experimental diesel engine.
Magnification × 140,000.

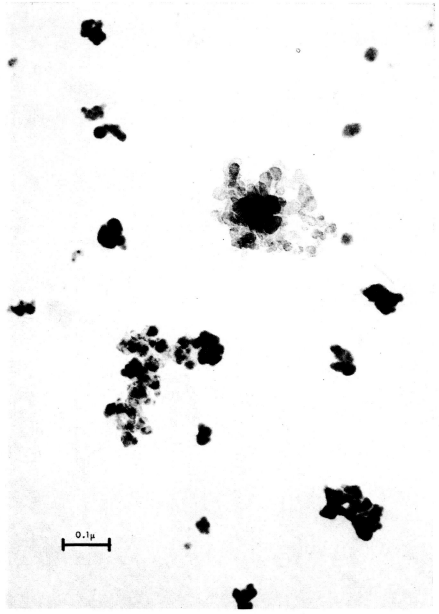

FIG 5. Particles from the diluted exhaust of a 1300-cc car engine using leaded petrol.
Magnification × 140,000.

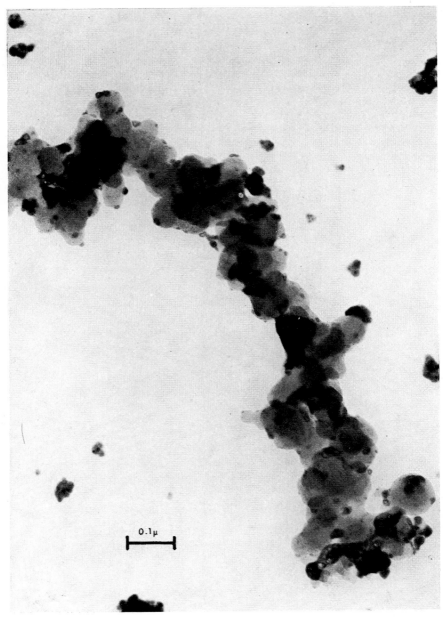

FIG 6. Aggregate from same source as Fig 2. Magnification × 140,000.

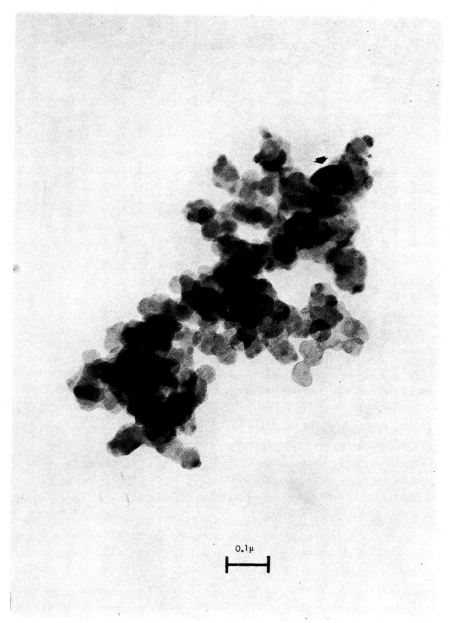

0.1μ

FIG 7. Particles from the general atmosphere in Fleet Street. Magnification × 140,000.

are mainly on the surface of less dense, and much larger, spheres but in some instances are apparently incorporated in smaller aggregates of smaller ultimate particles: the diameter of most of these dense areas is less than 0·015 μm, and some appear less dense than others. There are no such dense areas on the micrograph of diesel smoke, on which the absorption of the electron beam is greatest where its path through the aggregate is longest. It is not possible to state what these dark spots are simply on the basis of their appearance under the electron microscope, but it is reasonable to suggest that they contain part of the lead originally present in the petrol as alkyl. Similarly, the more solid appearance of the spherical constituent particles may plausibly be attributed to lead being distributed, as a dilute solid solution, within them. In the discussion below we suggest physical mechanisms by which these two types of particles might be formed. The aggregate shown in the micrograph of the sample from Fleet Street (Fig 7) shows even more clearly than those of Fig 4 that in many parts of the image the electron density is simply an indication of the thickness of the specimen at that point. At one or two places there are much larger dense areas, such as that indicated by the arrow: in general appearance these dense areas are very similar to those areas of the sample studied by microprobe analysis that showed the presence of lead, but it is difficult to say how they came to be there as they seem too big to have condensed directly from the vapour phase and rather too small to be likely to have been blown off the exhaust system.

Electron microprobe analysis: A sample collected in Fleet Street was examined for lead. It was found in the electron dense regions found in most of the larger aggregates (see above) but the fainter smoke particles did not show any more lead than did the background. There were few isolated single particles but these were not electron dense and showed no lead. The very small, electron-dense, areas found in the petrol exhaust fumes under the ordinary electron microscope would be too small to be picked out by microprobe analysis, since the diameter of the microprobe beam is about 0·2 μm and thus over ten times that of the particles.

DISCUSSION

Concentration of lead in ambient air: Although alkyl lead vapour is present in town air and may constitute a health hazard, none of our measurements or observations relate to it and it will not be considered here. So far, we have confined our studies to the particulate lead but our method of analysis might not detect all of the lead in organic lead compounds.

Many workers have published measurements of airborne particulate lead, but in general the figures quoted are not comparable to one another because the measured concentrations are critically dependent on the sampling site, on the time of day when the sample was taken, and on the length of time of sampling. Thus, close to freeways in Los Angeles "average" concentrations up to 30 μg/m^3 are not uncommon and for periods of two to three hours figures up to 71 μg/m^3 have been measured.[2] In the U.K. average daytime concentrations in streets over a long period (*e.g.* three months) of up to 5 μg/m^3 have been reported[1] and similar concentrations have been found in other European countries.[12] Much of the airborne lead in streets comes from the exhaust of petrol engines and concentrations away from traffic are considerably lower[1, 2, 13]; in our 1962–63 study we found the airborne lead 50 m from Fleet Street to be only about a quarter of that on the street itself. Moreover, most people spend much of their time indoors and concentrations both of total particulate matter and of lead[14] are commonly lower indoors than out. From the data now available it is therefore almost impossible to estimate with reasonable accuracy how much lead a particular individual inhales.

In the absence of established "norms" for town air, our measurements in 1971 can be assessed only by comparison with measurements at the same sites in other years, and these comparisons are so dependent on the vagaries of the weather during the periods compared that they provide a flimsy basis for constructive discussion. Moreover, the results we have obtained are equivocal, since in Fleet Street between 1962 and 1971 the lead concentration in the air appears to have risen substantially, whereas on the roof of our laboratory what slight trend there has been since 1965 appears to be downwards. A possible explanation of this discrepancy is that the contribution of sources other than petrol engines has been decreasing at the same time as that due to petrol engines has increased, for comparison of the "dilution factors" R for the various pollutants appears to indicate that a very considerable proportion of airborne lead at our roof sampling house does not come from vehicles. However, since the comparison of concentration in Fleet Street is between the years 1962 and 1971 and that on our roof is for the years since 1965, it is conceivable that the apparently different trends are due to chance fluctuations of the time series.

Blood lead in taxi-drivers: It is not necessary to go into a detailed account of the fate of lead absorbed in the body, which has been studied extensively. Here it need only be said that lead in the body as a whole, and in the blood, is not permanently fixed: rather there is a dynamic exchange involving continuous turn-over and, in particular, the blood is essentially a means of translocation for the lead rather than a permanent reservoir. Little is known of the uptake of lead from the lungs, although it seems likely that, from particles of the size found in the air, what absorption there is will not take longer than a few days at most: that through the gut cannot take substantially longer than two or three days. What is in the blood may, however, remain there for considerably longer before being excreted, laid down in the bone, or incorporated in other tissue. Consequently, blood lead concentration probably reflects intake over a period longer than does carboxyhaemoglobin. To this extent, therefore, the latter, which we have used as a measure of exposure to traffic, is not strictly comparable: haemoglobin reaches equilibrium with ambient concentration of carbon monoxide over a period of the order of 12 hours, and in the taxi-drivers HbCO saturation therefore reflects only exposure during the shift. Nevertheless, despite the limitations of the techniques used and the paucity of subjects in the study, the similarity of the figures of blood lead for day workers and night workers appears to indicate that uptake of lead from traffic fumes does not make a major addition to the body burden of the metal.

Electron microscope studies: The electron microscope studies offer more scope for comment. For many years industrial hygienists have described particles of industrial dust in terms of an "aerodynamic diameter" which is that of a sphere of unit density and the same falling speed in still air. This is appropriate in considering the particles that cause the classical pneumoconioses, because these particles are mainly of such a size, shape, and density that their deposition in the lung is principally due to sedimentation (and also, to some extent, to impingement, which depends on the same parameter). It has long been known, however, that, for particles substantially smaller than $0·5$ μm "aerodynamic diameter", the falling speed was so low that diffusion under Brownian motion became a more important means of translocation than settlement under gravity: present-day ideas on the subject date back to Findeisen's paper published in 1935, and although there is dispute about the exact size at which diffusion becomes more important than settling, there seems no reason to doubt the broad conclusions drawn from Findeisen's work (see, for example, the ICRP Task Group on Lung Dynamics report on "Deposition and Retention Models for Internal Dosimetry of the Human Respiratory Tract".[15]

Despite this, much of the discussion of lead aerosols in car exhausts has been based on the concept of the aerodynamic diameter of the particles (as in "Appendix A—Particle Size" in the U.S. National Research Council's report entitled "Airborne Lead in Perspective")[2]; the figures quoted for the particle size distributions of these aerosols have been derived from samples collected either by Goetz aerosol spectrometer or by Andersen impactor, both of which select in terms of what is, in effect, falling velocity. These sampling devices are able to rely on inertial forces because they magnify them (the former centrifugally, the latter by jet impaction) while leaving diffusion unchanged, so rendering largely ineffective the mechanism relevant to the problem. To get the relevant equivalent diameter it would be necessary to use a method of sampling which relied on diffusion to transport the particles towards the collecting surface. Only if the particles were well-defined in shape and density would it be possible to convert the "aerodynamic diameter" into one relevant to diffusion.

The electron micrographs (Figs 1–7) show that the shapes of the particles are far from well-defined. The particles differ from those for which the concept of "aerodynamic diameter" was devised in every relevant respect: they are notably smaller, all except the smallest are aggregates of extreme and diverse shape, and the effective density of these aggregates for aerodynamic purposes is quite unknown. To this it may be added that their behaviour with respect to diffusion is likewise unknown: it seems possible, however, that their Brownian diffusion rates will be lower than those of compact particles of the same volume because Brownian diffusion arises from the statistical fluctuations of molecular bombardment and the greater surface area of the aggregates will make these statistical fluctuations relatively smaller. Moreover, it can be said with reasonable certainty that diffusion will be the most important mechanism by which those particles that reach the surface of the lung are propelled there, and that their size and state of aggregation will have an influence both on whether or not they are deposited and on where they are deposited if they are. The form of the particle may affect the accessibility of the lead to the body fluids that will leach it away (since the lead compounds reported to be in car exhaust gases are too soluble to remain undissolved for long if water can reach them), and the site of deposition will determine whether uptake is through the respiratory tract, from which almost all will go into the blood, or through the gut, from which absorption is much less complete.

It is clear from the above that it is not possible to derive a "mass mean equivalent" parameter that has any meaning. It would have to be of the form:

$$[\Sigma_{\substack{\text{all} \\ \text{particles}}} \text{(soluble lead content of particle)}$$

$$\times \{ \int \{(\text{probability of deposition on area } \delta a \text{ of surface of the respiratory tract})$$

$$\times (\text{Efficiency of uptake of lead from particle deposited at } \delta a \} \times \text{d}a \}]$$

$$\div \Sigma_{\substack{\text{all} \\ \text{particles}}} \text{(total lead content of particle)}$$

(the integral being over the surface of the respiratory tract), which is clearly impractical. Any future studies must take as their starting point the acceptance of this inconvenient fact.

It is surprising that the above aspects of the problem do not appear to have been studied before. As far as the shape and composition of particles is concerned, on theoretical grounds one would expect most of them to be aggregates of smaller ones and to contain many and diverse chemical components. Lead in petrol is very dilute and on a rough calculation it would take every lead atom in a cube of side 80 μm to form a spherical particle of PbClBr of diameter 0·05 μm, and particles of this size cannot therefore comprise lead compounds only. The time available for particle formation is short, so that the primary particles must be small, and since they are small they will diffuse and aggregate rapidly. If lead-containing particles are formed first they will subsequently attach themselves to other particles, which is how we explain the small dense areas on the surface of some of the particles in Figs 5 and 6, whereas if they condense out at the same time as the other particles they will be incorporated in dilute form in particles that mainly consist of other elements, which is how we explain the more "solid" appearance of the particles in petrol exhaust.

In view of the various uncertainties about the details of deposition in the lung, clearly it is not possible to quote a figure for the fraction of inhaled lead that is deposited in the various parts of the respiratory tract, but if the diffusion rates of the particles present are lower than those of compact particles of the same volume, as is suggested above, deposition will be lower than that of spheres of the same mass and volume. There is no general agreement as to what this figure should be. The most reliable data on deposition of inhaled particles are probably those of Davies and Muir, summarized in ref. 16, using spheres of diameter 0·5 μm. Deposition depends on tidal volume and breathing rate, as well as on particle size, but for values of these that correspond to light activity these authors found deposition was about 10–12 per cent (Fig 1, ref 16). These figures are dramatically lower than those suggested by the ICRP Task Group for particles of the same size,[15] and since the experiments of Davies and Muir were carefully planned and meticulously executed, we are forced to conclude that the ICRP figure for particles of this size and density is too high by a factor of at least 2, and probably more. Taking everything into consideration, we find it difficult to avoid the conclusion that the same is true of the smaller particles relevant to the present problem. Using the figures indicated in Fig 11 of the Task Group's report, we get about 60 per cent pulmonary deposition and 25 per cent tracheo-bronchial deposition. The former figure seems inordinately high, and any reduction of it is of immediate relevance to total uptake, since it must be assumed, until proved otherwise, that all lead deposited in the non-ciliated regions of the lung is taken up by the body. The smallest particles diffuse very rapidly and for these the tracheo-bronchial deposition, on ciliated areas of the respiratory tract, will to some extent protect the lower parts of the lung, which in turn will reduce total uptake by the body because much of what is deposited on ciliated surfaces is likely to be removed by ciliary action and swallowed (so that absorption is that characteristic of the gut). Moreover, the Task Group's Fig 11 refers to a tidal volume of 1450 ml, which corresponds to moderate to severe exercise and for the smaller tidal volume characteristic of breathing at resting level the deposition will be lower. It must be stressed, however, that for particles in the size-range of those seen in the electron micrographs there are no data on deposition that compare in reliability with those of Davies and Muir for particles of 0·5 μm.

In conclusion, it would appear that little of the speculation on the uptake of lead inhaled in the form of aerosols in the exhaust from petrol engines is based on solid and established fact. There are numerous measurements of airborne concentrations of lead, but this is very variable in time and space and the lead burden imposed by breathing is much more difficult to assess than that imposed by drinking water from lead pipes. Moreover, the particles we see in our electron micrographs are very different from those of lead sesquioxide used by Kehoe,[17] or of lead fume used by Nozaki,[18] and there does not appear to be available any detailed study of the deposition in the respiratory tract of the particles, containing lead, from vehicle exhausts. In the absence of such data, the only evidence relating to the effect of these exhaust gases is from epidemiological studies on man. If inhalation of air contaminated by lead from car exhaust were a major source of the body burden of lead, one ought to see a definite relationship between blood lead and airborne lead. An urban-rural gradient in airborne lead concentration has been demonstrated, though the contribution made by the motor vehicle is still in doubt, but there is not a commensurate gradient in blood lead.[19, 20] The study of a small population of taxi-drivers does not support the hypothesis that the main source of blood lead is traffic. These results and those of other workers indicate the need to extend work on the sources and biological availability of airborne lead and on the variation in body burden with exposure to it.

ACKNOWLEDGMENTS

The authors would like to express their thanks to Jean Peal, Leslie Hampton, and Nicholas Kollerstrom, who did the chemical analysis; to Cyril Brown, who arranged the engine experiments; to Mrs Anita Furzey, of the National Physical Laboratory, who did the electron microprobe analysis; and to the Corporation of the City of London for their helpfulness in providing space for sampling equipment in Fleet Street.

REFERENCES

1. Waller, R. E., Commins, B. T., and Lawther, P. J. *Brit. J. ind. Med.*, 1965, **22**, 128–38.
2. A report prepared by the Committee on Biological Effects of Atmospheric Pollutants of the Division of Medical Science. National Research Council National Academy of Sciences. Washington D.C., 1971.
3. Laveskog, A. Paper presented at Second International Clean Air Congress, Washington. 1970, Paper CP-37D.
4. Commins, B. T. *Atmos. Envir.*, 1969, **3**, 565–72.
5. Waller, R. E., and Commins, B. T. *Envir. Res.*, 1967, **1**, 295–306.

6. Saltzman, B. E. *Analyt. Chem.*, 1954, **26**, 1949.
7. Commins, B. T. *Analyst*, 1958, **83**, 386–89.
8. Stocks, P., Commins, B. T., and Aubrey, K. V. *Int. J. air & wat. Poll.*, 1961, **4**, 141–53.
9. "Lead in Gasoline—U.K. Levels." *Petrol. Rev.*, 1971, 289.
10. Waller, R. E., Commins, B. T., and Lawther, P. J. *Brit. J. ind. Med.*, 1961, **18**, 250–59.
11. Jones, R. D., and Commins, B. T. To be published.
12. Blokker, P. C. "Health Aspects of Lead Emissions from Gasoline Engines." 1970, Stichting CONCAWE, The Hague.
13. Daines, R. H., Molto, H., and Chilko, D. M. *Env. Sci. & Tech.*, 1970, **4**, 318–22.
14. Yocom, J. E., Clink, W. L., and Cote, W. A. *J. air poll. Control Ass.*, 1971, **21**, 251–59.
15. ICRP Task Group on Lung Dynamics. *Health Phys.*, 1966, **12**, 173–207.
16 Davies, C. N. "Aerosol Sampling Related to Inhalation". International Atomic Energy Agency, Vienna, Paper No. SM-95/1, 1967.
17. Kehoe, R. A. *J. air poll. Control Ass.*, 1969, **19**, 690–701.
18. Nozaki, K. *Ind. Health (Japan)*, 1966, **4**, 118.
19. Ludwig, J. H., Diggs, D. R., Hesselberg, H. E., and Maga, J. A. *Amer. ind. Hyg. Assoc. J.*, 1965, **26**, 270–84.
20. Tepper, L. B. Seven-City Study of Air and Population Lead Levels: an interim report, 1971. Department of Environmental Health, College of Medicine, University of Cincinnati.

Discussion

Dr A. Berlin (Commission of the European Communities): Could you please indicate the exact position with respect to traffic of the sampling point on Fleet Street? What is the sampling efficiency of the glass fibre filters and is there any lead present originally in the filters?

Dr Egan reports in his paper the blood lead concentration data obtained by Kubota in the U.S.A., 13·2 μg/100 ml. To what do you attribute the difference between this value and the 26·29 μg/100 ml average values which you measured in taxi-drivers?

Dr McCallum indicates in his paper that estimations of ALA and ALAD are likely to be more sensitive indicators of exposure to lead. Could you tell us why ALA and ALAD analysis were not carried out at the same time as blood-lead analysis?

Professor P. J. Lawther: The sampling point in Fleet Street was in the middle of the road at mouth height. The sampling efficiency of the glass fibre filters is virtually 100 per cent and there is no lead present originally in the filters, though blanks are done. The blood figures collected in the U.S.A. (13·2 μg/100 ml) refer to only one set of many series of measurements. There are other series which are quite in line with the ones we reported. We intend to do ALA and ALAD measurements, but they do not replace blood lead.

Professor D. Bryce-Smith (University of Reading): Professor Lawther states that the average lead levels in Fleet Street increased by about 68 per cent from 1962 to 1971, but that "the number of vehicles involved is unlikely to have risen substantially". Yet according to the GLC, the traffic levels in Fleet Street were 22,300–26,000 vehicles/day in 1962 and 30,000–36,000 in 1970, an increase of nearly 40 per cent. To what extent can measurements in Fleet Street provide guidance for other locations, *e.g.* the Cromwell Road extension, which carries 116,000 vehicles/day?

Professor P. J. Lawther: Pollution by any substance emitted by cars is not necessarily proportional to the number of vehicles per hour passing the sampler. Obviously, pollution may be highest when there are no vehicles per hour passing the point, *i.e.* when there is a traffic jam and all the vehicles are idling. The measurements in Fleet Street can give only a rough indication of pollution of busy traffic; measurements at other sites are desirable.

Dr R. M. Hicks (Middlesex Hospital Medical School): What is the pore size of filters for air samples, in view of EM findings that the mean diameter of particulate matter is less than 0·01 μg.

I query whether the dense particulate matter in EM, referred to by the speaker, really are lead. EMMA is only 0·2 μg beam size. The arrowed particle in Fig 7 is much, much smaller than this. The method shows that the whole sample contains lead, but not where it is in the particles shown. The method shows that the whole sample contains lead, but not where it is in the particles shown. The dense particles could be any other metal.

Professor P. J. Lawther: Tests of filter efficiency have shown that no lead gets through the filters we have used in our analysis. Dr Hicks was right in saying that the particle pointed out on the screen was less than 0·2 μg. The electron probe analysis took in most of the aggregate in which this smaller dense particle was found. It is possible that the dense particles could be any other metal but it would seem reasonable to assume that they were lead because of their number.

Lead Uptake by Plants and its Significance for Animals

By L. H. P. JONES and C. R. CLEMENT

(The Grassland Research Institute, Hurley, Berkshire)

INTRODUCTION

IN introducing our subject, it would be helpful to say something about the mineral nutrition of plants. At least 16 elements are essential for the healthy growth and development of higher plants; they are as follows: carbon, hydrogen, oxygen, nitrogen, phosphorus, potassium, calcium, magnesium, sulphur, iron, manganese, copper, zinc, boron, molybdenum, chlorine. A seventeenth, cobalt, is needed for legumes, at least, when fixing their own nitrogen. Carbon, hydrogen, and oxygen are derived from the air and from water. Carbon, as carbon dioxide, and some oxygen enter the leaves through minute pores, or stomata, in their surfaces; additional oxygen and hydrogen enter as water through the roots. The other 13 (or 14) elements are supplied by the soil, and enter the roots as ions dissolved in the soil water. In addition to the essential elements, the soil water also contains other elements which enter the roots. These include trace elements that are required by animals, such as selenium and fluorine, and others, such as lead, that are not at present known to be required either by animals or plants.

Because of the nutritional hazard of lead to animals and man, agricultural scientists have been concerned with understanding (i) the soil properties and processes that determine the supply to roots; (ii) the plant processes of uptake and transport from roots to shoots; and (iii) the plant and animal factors that determine intake and absorption by animals. In the present paper we shall consider each of these aspects of the movement of lead from soil through plant to ruminant animals and, finally, to man.

LEAD IN SOILS

Lead is a natural constituent of the lithosphere, in which its abundance is 16 ppm, compared with 70 and 80 ppm for the essential trace elements copper and zinc, respectively.[1] The total content of lead in agricultural soils is also of the same order as that of copper and zinc, and usually lies between 2 and 200 ppm.[2] The content in some agricultural soils of Scotland examined by Mitchell and co-workers[3, 4] and in those which we have examined[5] from England and Wales extend over this range (Table I.)

We shall now consider, in its simplest form, the supply of lead from soil to plant roots. The chemical form and reactions of lead in soils are still very imper-

TABLE I

Lead Content of some Agricultural Soils

Location	Number	Total Pb (ppm)	HAc (ppm)	EDTA (ppm)	Reference
Scotland	2	30–50	1·1–1·6	7·0–12·0*	(3)
Scotland	3	20–70	0·24–0·7		(4)
Wales	2	75–205	0·64–2·03	22·0–59·0†	(5)
England	14	20–149		5·3–41·0†	(5)

* Extractant—EDTA neutralized with NH₃.
† Extractant—EDTA in NH4Ac.

fectly understood, but it is known that only a very small proportion of the total amount in a soil is present in solution; in this respect, lead resembles many other ions. While a plant is growing, its roots remove ions from the soil solution and these are replenished from the solid phase. The rate of replacement largely determines the supply to the plant, and laboratory procedures to estimate this usually involve extracting the ion in question from the soil with various reagents. Solutions of acetic acid and of ethylenediamine-tetra-acetic acid (EDTA) have been used in studies of lead in U.K. soils and some data from these are given in Table I. Although the latter reagent extracts about a quarter of the total soil lead, it should not be assumed that all of this is available to the root in the short term.

UPTAKE AND TRANSPORT IN GRASS PLANTS

In the strictest sense the supply of available lead is that actually extracted by plants from the soil, and we have determined this in the range of 16 agricultural soils from England and Wales referred to in Table I. The experimental plant, perennial ryegrass, was grown on these soils in a controlled environment cabinet with filtered air, day/night temperatures of 23°/18° C, light intensity of 11 × 10⁴ ergs/cm²/sec, and a 16-hr photoperiod. The ryegrass was harvested

TABLE II

Lead Uptake by Perennial Ryegrass Grown on Soils in Controlled Environment Experiment[5]

Location (County)	Extractable Pb‡ (ppm)	Roots (ppm)	Roots (μg)	Shoots (ppm)	Shoots (μg)
Nottingham	5·3	10·0	29·8	5·1	17·4
Northumberland	6·9	11·9	44·1	5·2	29·7
Lancashire	40·8	36·8	90·5	6·0	33·1
Cardigan	59·0	37·8	110·0	7·4	27·3

* From Experimental Husbandry Farms of the Agricultural Development & Advisory Service, Ministry of Agriculture, Fisheries & Food.
† Grown with added K, Ca, Mg, N, P, S.
‡ With NH4Ac plus EDTA.

by cutting at about 2 cm above the soil surface on four occasions, and at the final harvest the roots were separated from the soil and analysed for lead. The content in the roots of plants grown on four of the soils is shown in Table II, together with the content in the shoots from the final harvest; the shoots from preceding harvests had similar contents. The main point to be noted is that although the total content of lead in roots plus shoots increases with increasing levels in the soil, this is largely due to accumulation in the roots.

In order to study this accumulation in the roots, and the related processes of uptake and transport from roots to shoots, it is necessary to dispense with the complex soil system and to grow plants in solution-culture in which the essential elements are supplied as simple salts in water. Accordingly, we have grown perennial ryegrass in solution-culture, and at an appropriate stage of growth we added lead (as Pb $(NO_3)_2$) to the solution. After three days the solution was changed to one that was free of lead, and at intervals thereafter we removed plants and determined lead uptake and transport to the shoots by analysing the plant material. Table III gives the results from two experiments. The main

TABLE III

Lead Uptake by Perennial Ryegrass Grown in Solution-Culture[5]

Level of Pb added to solution		Pb content of plants at 14 days after addition to solution			
		Roots		Shoots	
(ppm)	(μg)	(ppm)	(μg)	(ppm)	(μg)
0·4*	400	98	201	3	37
1·0†	12500	3696	9610	104	1055

* In 1 l of static solution-culture; controlled environment experiment.
† In 12·5 l of flowing solution-culture; glass-house experiment.

points to be noted are (i) with 400 μg Pb (0·4 ppm) in solution, the concentration in the shoots was similar to that in the soil-grown plants (Table II); (ii) with 12,500 μg Pb (1·0 ppm) in solution the content in the shoots increased from 3 to 104 ppm, i.e. by a factor of 35, whereas the content in the roots increased by a factor of 40.

The experiments with solution-cultures show that ryegrass roots take up lead readily but pass on only a small proportion to the shoots. In this way they provide a barrier which restricts the movement of lead from soil through plant to ruminant animals.

These remarks apply to actively growing ryegrass plants, but during the winter period, when the rate of growth is low, considerable increases have been observed in the lead content of the shoots. For example, Mitchell and Reith[3] report increases from 0·8 and 1·5 ppm in August/September to 5·1, 11·9, and 21·8 ppm in late November in ryegrass, cocksfoot, and mixed pasture respectively. However, this apparent increase in the transport of lead from roots to shoots was confined to the months of restricted growth and in the succeeding spring the content of lead in the mixed pasture fell to 5·7 ppm in April and to 2·1 ppm in May.

INTAKE AND ABSORPTION IN RUMINANT ANIMALS

Soil to root to shoot is the major pathway of entry of lead, and of most other trace elements, into both plants and animals. Lead ingested by animals is poorly absorbed in the alimentary tract, and Blaxter[6] has found in sheep that 98·7 per cent of daily intakes ranging from 1·9 to 115·5 mg Pb was excreted in the faeces, thus giving an apparent absorption of only 1·3 ± 0·8 per cent. It may be noted that these intakes would be equivalent to 1·9 to 115·5 ppm Pb in the dry-matter intake (typically 1 kg daily) of a sheep of 50 kg live weight. The low absorption of lead can be regarded as a barrier which further, and substantially, restricts its movement into the tissues of the animal. In broad terms, the lead which is absorbed is partly excreted in the urine and via the bile in the faeces, and partly retained in the bones and soft tissues of the body. Blaxter[6] has shown that in sheep with daily intakes up to 3 mg the total excretion keeps pace with ingestion and there is no retention of lead. Above this level of intake, lead is retained in increasing quantities by the tissues, particularly in bone. The affinity of bone for lead is apparent from the observation by Schroeder and Tipton[7] that 91 per cent of the total body burden in adult man is present in the skeleton. Blaxter[6] has suggested a similar partitioning in the body of sheep. Critical levels of intake appear to be higher in ruminant animals than in man. Thus the lower limit of safety for man has been given by Cantarrow and Trumper[8] as 0·2 to 2·0 mg Pb daily, and for non-pregnant ewes as 6 mg/kg live weight[9, 10] which is equivalent to 300 ppm Pb in the dry diet.

CONCLUSIONS

Although there are many gaps in our knowledge of the chemistry of lead in soils and the mechanisms of uptake and transport in plants, it may be concluded that there are four barriers restricting the movement of lead from soil to man at the end of the food chain. The first of these is in the soil, where only a small proportion of the total lead is available to plant roots. The second is in the roots, from which only a small proportion of the absorbed lead is transported to the shoots. The third is in the alimentary tract of the ruminant animal, from which only a very small proportion of the amounts ingested are absorbed. Finally, the fourth barrier results from the partitioning of lead in the tissues of the ruminant animal where it accumulates in the bones, and does not therefore contribute significantly to man's dietary intake.

REFERENCES

1. Goldschmidt, V. M. "Geochemistry". Oxford University Press, 1958.
2. Swaine, D. J. *Commonw. Bur. Soil Sci., Tech. Comm.*, 1955, 48.
3. Mitchell, R. L., and Reith, J. S. *J. Sci. fd Agric.*, 1966, **17**, 437–40.
4. Swaine, D. J., and Mitchell, R. L. *J. soil Sci.*, 1960, **11**, 347–68.
5. Jones, L. H. P., Clement, C. R., and Jarvis, S. C. Unpublished.
6. Blaxter, K. L. *J. comp. Path. Ther.*, 1950, **60**, 140–59.
7. Schroeder, H. A., and Tipton, I. H. *Archs envir. Hlth*, 1968, **17**, 965–78.
8. Cantarrow, A., and Trumper, M. "Lead Poisoning". Baltimore, Williams & Wilkins, 1944.
9. Allcroft, R., and Blaxter, K. L. *J. comp. Path. Ther.*, 1950, **60**, 209–18.
10. "The Nutrient Requirements of Farm Livestock, No. 2 Ruminants". Agricultural Research Council. HMSO, London, 1965.

Discussion

Dr P. S. Elias (Department of Health and Social Security): What is the uptake of lead by root vegetables and lettuce grown on allotments and thus by-passing the animal barrier? Are any figures available?

Dr L. H. P. Jones and Dr C. R. Clement (written answer): The most definitive work on lead in vegetable crop plants is that of Dr H. L. Motto and colleagues (*Environ. Sci. Technol.*, 1970, **4**, 231–37) in the U.S.A., who examined uptake and transport of lead in potato, lettuce, tomato, and carrot grown in sand culture. Such culture is similar to that employed in our experiments in that lead was added to a nutrient solution, thus avoiding the complexity of the soil. With potato, lettuce, and tomato, the roots contained much larger concentrations of lead than did the tubers, the leaves, or the fruit. This agrees with our findings with ryegrass and supports the concept of a barrier restricting the movement of lead from roots to shoots. There is, however, an apparent anomaly with carrot in which the concentration in the tops (leaves and stems) was similar to that in the roots. Indeed, Le Riche (*J. Agric. Sci. Camb.*, 1968, **71**, 205–8), working with carrot and beet, found a greater concentration in the tops than in the roots. It should be noted that in plants with storage roots, such as carrot, the lead that is absorbed from the soil is greatly diluted by the transfer of assimilates from the leaves.

When plants are exposed to aerial contamination, these inherent physiological effects may be masked. For example, Motto and colleagues report that at 10 m from a highway in the U.S.A. (traffic volume 47,100 vehicles/24 hr) the leaves and roots of lettuce both contained 24 ppm Pb, and those of carrot contained 37 and 6 ppm, respectively.

Trace Lead in Food

By HAROLD EGAN

(*Laboratory of the Government Chemist, Department of Trade and Industry*)

LAWTHER[1] *et al.* have already reviewed the occurrence of traces of lead in the atmosphere and have commented on the contribution of this to the total intake by man. The purpose of the present paper is to complete the picture by summarizing evidence relating to the occurrence of traces of lead in food and beverages with an account, so far as this is possible, of the origin of these traces and their contribution to the total intake by man. Goldschmidt[54] has estimated the average lead content of the entire earth's crust to be of the order of 16 mg/kg. So far as soil is concerned, Swaine[2] quotes Vinograder, who in 1938 gave the average lead content as 0·1 ppm Swaine summarized the literature on lead in soils to 1953, levels in Britain ranging mainly from 2 to 100 mg/kg. Much of this, however, is not in soluble form. So far as Britain is concerned, naturally occurring lead is or has been of commercial importance in many areas, several of which have been worked since Roman times. For the most part they are confined to the geologically older upland areas. In England the most important are in the North Pennines, notably in Cumberland (Alston Moor), Northumberland, Westmorland, Durham (Upper Teesdale and Weardale), and Yorkshire: in Derbyshire, chiefly in the Carboniferous Limestone; in many areas of the Lake District; in the Shelve-Habberley district of West Shropshire; in the Mendip Hills; and in Cornwall and Devon. In Scotland, the Leadhills district of Lanarkshire, Wanlockhead in Dumfriesshire, and Strontian and Tyndrum in Argyllshire are all famous mining areas. Wales has derelict lead mines in every county, the main areas of mineralization being in Flintshire, West Denbighshire, North Cardiganshire, West Montgomeryshire, and Anglesey. Lead occurs in Northern Ireland near Keady in County Armagh and near Newtownards in County Down, whilst the Isle of Man also had important mines at Snaefell, Foxdale, and Laxey.

LEAD IN WATER AND WATER SUPPLIES

The natural lead content of surface waters is of the order of 5 micrograms/litre: no water put into supply by a statutory undertaking in Britain contains more than the recommended limit of 50 micrograms/litre set by the World Health Organization.[3] Ettinger[4] has survey data on the lead content of water in the U.S.A., concluding that recreational boating contributed 400,000 kg lead from gasoline in 1963 as compared with the 8 million kg estimated by Patterson[5] to be contributed by sewage (some 50 per cent of which is probably removed during the course of sewage treatment). Surface waters were found to vary from 10 up to occasionally to 100 micrograms/litre Pb for samples passing an 0·45 micron Millipore filter (*i.e.* soluble lead), all but the most highly mineralized conforming to the WHO standard. Most drinking water supplies contain less than 10 micrograms/litre Pb. The principal hazard in Britain arises with private, untreated supplies, as illustrated in the 1967 Report of the Chief Medical Officer of the Department of Health.[6] Plumbosolvency may still be a consideration for soft waters, if not adjusted for pH to correct this, or for certain hard water with a high organic and nitrate content, since some older properties in England and Wales still have lead piping. Aggressive waters from moorland areas when piped may result in values as high as 25 mg/litre of lead when drawn after standing for several hours, e.g. from a bathroom tap in the early morning. WHO European Standards for Drinking Water[7] lay down a maximum lead concentration of 300 micrograms/litre after 16 hr contact with the pipes. An untreated private supply with the maximum amount of lead allowable would, on the assumption that 10 per cent of lead is absorbed and that two early morning cups of tea are consumed, give a daily uptake of about 20 micrograms. Lead salts may be used as stabilizers in plastic pipes. Packham[8] has reviewed the possibility of leaching occurring in this circumstance, finding that the lead concentrations in water samples from unplasticized polyvinyl chloride distribution systems were in general very low; however, water which was undisturbed in newly installed pipes for several hours initially contained 50 micrograms/litre, falling to acceptable levels with usage within a few days or weeks.

Trace lead content of ice has been measured at various depths, representative of different periods of time, by Murozumi et al.[9] in Greenland and by Jaworowski[10] in Poland. The levels concerned varied from about 0·02 to about 0·2 micrograms/kg. The increases found with time were related to increased general industrial activity, including the volatilization of lead during fuel combustion, though some doubt was cast on the relationship suggested by the former workers by Mills[11] on account of possible contamination of a remote site by the transport used to gain access.

LEAD IN SOIL AND VEGETATION

The effects of air pollution on plants and soils was considered in relation to lead by the Agricultural Research Council in 1967[12], with the conclusion that the limited data available did not provide evidence of significant contamination of the soil near roads by exhaust fumes. Lead is normally present in soils to the extent of 10 to 100 mg/kg, with more in the surface soil layer than in the lower horizons: accumulations in top soils are normally due to the presence of lead in an insoluble form in plant residues. Plant species vary considerably in their "normal" lead contents, there being little reliable information on the relationship between plant uptake and the amount of acetic acid extractable lead in the soil. In Britain, lead arsenate is rarely used as an insecticide, and only on apples and pears, with a minimum interval of six weeks between application and harvest, or on non-edible crops.

For vegetation near busy highways in Britain, however, the evidence does suggest some contamination by lead derived either directly from exhaust fumes or from contaminated dust. The effect is largely one of surface deposition, with the outer leaves (of cabbage, for example) showing more lead than the inner leaves. Thus for crops growing near the Great North Road in 1964 the lead content (ppm oven dry material) fell from 5·0 at 10 yards to 1·0 at 150 yards from the highway for outer leaves and from 1·0 to 0·4 for inner leaves. Jones and Hatch[13] compared various species growing on treated and untreated soils, the comparative levels being about 11 mg/kg and 7 mg/kg respectively. Donovan et al.[14] studied pastures in relation to lead mining operations in the West of Ireland, finding the lead content of grass on adjoining land to vary to up to well over 2000 ppm, whereas samples from remote sites were very much lower (3 to 35 ppm). Similar results were found for roadside dust samples. Goodman[15] has studied the re-vegetation of derelict land contaminated with toxic heavy metals and quotes a government survey in 1963 in which it was estimated that about 40,000 ha of land in England and Wales was so damaged by industrial use as to be incapable of further use without treatment. Much of this land (which represents about 0·25 per cent of the total land surface) is located around the larger centres of population. Apart from mining for iron, lead, and other heavy metal ores, these areas are associated with quarrying and mining for brick-earth, china-clay, building stone, limestone, coal, sand, and gravel. The accumulation of pulverized fuel ash from power stations, wastes from chemical plants, and household refuse all add to the disposal problem. Lead is not the dominant or most significant feature in all of these accumulations, nor is it easy to give a clear account of its relative importance in all cases. Rains[16] has demonstrated a cumulative increase in lead content of wild oats Avena fatua in an area where smelting activities had been established for many years. Levels in grown plants rose during ripening to up to 500 ppm lead on a dry weight basis. Mitchell and Reith[17] however, have suggested mobilization of lead by the root system of pasture herbage during senescence, with levels of 0·3 to 1·5 ppm during the period of active growth, rising to 30 to 40 ppm by the late winter or early spring. Maclean, Halstead, and Finn[18] studied the extractability of added lead from soils, finding the amounts taken up by oats and lucerne were inversely proportional to the organic matter content and to the pH of the soil. The uptake was reduced by the addition of phosphate or lime to acid soils. Markland and Vallance,[55] however, found little evidence of uptake by vegetables grown on composts to which lead had been added to simulate soils in the lead-bearing areas of Derbyshire.

In 1928–29 a Ministry of Health Departmental Committee on Ethyl Petrol[51] in Britain had various samples of settled dust taken from buildings in London and elsewhere analysed for lead at a time when ethyl petrol (as it was then called) was only just beginning to be used. It concluded that lead was universally present in the dust of London and the surrounding country towns and districts, the proportion varying from zero to 3·3 per cent by weight. Settled dust in garages was also examined, the average lead content (0·25 per cent) being similar to that of the dust on buildings.

The influence of atmospheric lead in the lead content of food and drinking water has been reviewed by Blokker.[56] In field experiments designed to assess the relative importance of air, water, and soil as sources of lead in ryegrass and radish crops, Dedolph et al.[57] found that 2–3 micrograms lead/g dry weight of leaf derived from the soil and that additional levels derived from and were quantitatively related to atmospheric concentrations of lead.

Motto et al.[19] sampled soils and plants from the proximity of busy highways in the U.S.A. and showed that the lead contents tended to increase with traffic volume and decrease with distance from the road. The lead content of unwashed grass samples varied from 133 ppm at the roadside to 34 ppm 75 yards away for a traffic volume of about 20,000 vehicles/hr with approximately twice these levels at 55,000 vehicles/hr and with somewhat similar results for soil. Most of the lead accumulation was found within 25 yards of the highways and most of the lead was found in the top 6 inches, the authors concluding that any downward movement is only slight. More lead was found in the upper parts of field-grown crops than on the lower parts. Lead tended to accumulate in the roots of green-house plants grown on sand in proportion to the amount of lead added: this effect also showed up in the leaves, indicating absorption of the lead by the roots with some translocation. Similar results have been found by Chow[20] and by other workers in Canada[18] and in Britain.[12, 21] In Switzerland, where Bovey[22] fed dairy cattle with hay contaminated by lead from car exhausts, much of the lead so ingested was soon eliminated, but after four weeks the lead content of milk increased fourfold. Some 140 mg of lead per animal per day was estimated to be absorbed from untreated hay containing 10 ppm lead, whereas some ten times this amount was ingested from roadside fodder containing 100 ppm lead. Indeed, Goodman and Roberts[23] have studied the levels of lead and other trace metals in plants and soils as indicators of these metals in the air. Brandt and Bentz,[58] in a co-operative analytical study, found lead in market whole milk to be of the order of 5 micrograms/litre.

LEAD IN FOOD

Trace metals such as copper, manganese, zinc, and cobalt are essential consti-tuents of man's diet. Lead is not. Patterson[5] has estimated that nearly 90 per cent of the lead ingested by man is derived from food, although only some 5 per cent of this is absorbed. Traces of lead in the diet are partly natural and partly technological in origin. An important technological source derives from the lead alkyls used as petrol additives, which is eventually transported by way of the atmosphere on to water and food crops. Other sources are lead in glazed pottery and (less so now than formerly) lead arsenate used as a pesticide on food crops and tobacco. In Britain the use of lead arsenate in agriculture is confined to orchard crops, where it has been largely replaced by synthetic organic pesticides. However, lead shot are occasionally found in game, up to 19 being counted by Hubbard and Pocklington in one commercial glass pack of pheasant. Bagley and Lock[52] examined bone and liver samples of wild birds, finding mean levels of the order of 3 to 13, 0·5 to 3·5 mg/kg lead (wet tissue) respectively.

The deliberate addition of lead chromate to improve the appearance of spices such as turmeric has long since been abandoned. Lead sulphide may still be used as an eye cosmetic in some communities.[24]

The average daily intake of lead by man in the U.S.A. was estimated by Kehoe[26] as being of the order of 400 micrograms; of this, some 300 micrograms was derived from food, with perhaps 50 micrograms from the atmosphere and 20 micrograms from water. Natural lead is ubiquitous; for this reason it is not normally possible to trace individual pathways by which traces of natural lead enter food.

The basic data on the occurrence of traces of lead in food in Britain were assembled in 1938 by Monier-Williams,[27] who listed the main ways by which food might become contaminated with lead as:

1. The use of alloys, solders, enamels, and glass containing lead, in plant and vessels used for manufacture, storage, transport, and cooking.
2. The use of ingredients such as citric, tartaric acid, acid calcium phosphate, and synthetic dyes in the production of which materials containing lead have been used or which have been manufactured in lead-lined vessels.
3. The use on fruit and vegetables of agricultural sprays containing lead arsenate or other lead compounds.
4. Exposure to dust containing lead such as that produced by the weathering of lead paint.
5. In the case of shell fish and crustacea, "natural" lead presumed to be extracted from sea water.

The ingredients such as citric acid to which he refers, and which in 1907 had been the subject of a special Local Government Board enquiry in Britain, are now the subject of rigorous purity specifications,[28] being today of pharmacopoeial quality as exemplified in the Food Chemicals Codex.[29] Dr Monier-Williams was a member of the Food Standards Committee and of its Metallic Contamination Sub-committee in Britain at the time of the Committee's original report on lead.[30] In 1954 the Committee reported that lead was one of the most widespread and serious of metallic contaminants in food and drink. Whilst many foods contained small amounts of "natural" lead, the principal source of contamination was the use of lead, lead alloys, and lead compounds in food processing plants, in the piping by which water and other fluids were conveyed, and from agricultural insecticides. It also noted that lead could be inhaled from atmospheric dust. Lead pipes on licensed premises and lead arsenate in orchard husbandry were the subject of special comment in relation to beer and cider, respectively, and in recommending limits for lead the Committee dealt separately with beverages, on the one hand, and other foods, on the other. When in 1961 for the first time regulations were made controlling the limits of lead in food,[31] a general limit of 0·2 ppm was scheduled for ready-to-drink non-alcoholic beverages and a special limit of 1·0 ppm for beer, cider, and perry to operate until 19 April 1964, with 0·5 ppm thereafter. For other foods a general limit of 2 ppm was set but with a number of specified exceptions, some of which were given a limit of 0·5 or 1·0 ppm and others a limit of 3·0 or 5·0 ppm, with canned meat at 10·0 ppm, reducing to 5 ppm in 1964. Fish and fish products were excepted to the extent that any lead content in excess of 2 ppm which was due to lead naturally present did not constitute an offence. All of these limits largely followed the earlier views of the Food Standards Committee, which also recommended that the use of lead piping for the conveyance of beer, cider, or other beverages, the use of lead-containing materials for packing or wrapping food, and the sale of domestic cooking equipment lined with lead-containing tin or pottery glaze should be controlled.

Contamination in the course of food processing is far less frequent than formerly, as indicated by Oser[32]: lead piping in food plants, breweries, and public houses and lead foil for lining tea chests have now been replaced by other materials. Metal equipment in the food industry is now constructed almost exclusively from stainless steel, with welded joints containing no lead, whilst similar changes have also occurred in home food preparation. The extent of leaching from soldered can seams is very small: Bergner and Miethke[33] found the lead uptake from fruit and vegetable liquors on storage was low in both lacquered and unlacquered cans. Even after storage for several years, few samples of canned corned beef have been found to exceed the statutory limit in Britain of 5 ppm.[34] That vigilance is still necessary is, however, illustrated by the finding in 1966 of alleged obsolete stocks of frying pans lined with alloys containing 40 to 60 per cent lead: a maximum of 0·25 per cent lead is allowed for the lining of cooking utensils by British Standard Specification 3788: 1964. Eggs, bacon, tomatoes, and sausages cooked in such pans removed from 6 mg reducing to 0·2 mg of lead per meal, some 65–70 mg being extracted by 100 meals.[35] Similar circumstances arise when fruit drinks or illicit liquor are made or stored in old lead-glazed pottery vessels. A ceramic dinnerware lead surveillance programme has recently been announced in the U.S.A.[53]

The view that lead has now largely been replaced by other materials in food processing and preparation is also supported by the Joint FAO/WHO Expert Committee on Food Additives,[36] which in 1966 expressed the view that inhalation from exhaust fumes and industrial smoke appeared to be increasing and that tobacco smoking might also contribute to lead intake. At the same time the use of lead arsenate as an insecticide appears to be decreasing: the annual production in the U.S.A. dropped from just over 7 million pounds in 1965 to 6 million pounds in 1967.[37] In 1956 it was 30 million pounds.[38]

Water distributed by the statutory undertakings in Britain complies with the World Health Organization's recommended European standard limit of 0·05 mg/litre (0·05 ppm), although overnight contact with domestic lead piping can give rise to higher values, as illustrated by Crawford and Norris.[39] Ullmann[40], in a continuing study in Connecticut, has reported little or no significant difference in lead in reservoir waters near to and remote from highways, most of the amounts found being less than 0·01 ppm. He also reported mean levels in shellfish of 0·16 mg/kg with 10 per cent of the samples examined exceeding 0·2 mg/kg, some of the beds (unrepresentative of the main culture beds) being in areas known to receive industrial waste.

The habit of some children of sucking or gnawing toys or other common objects, *pica*, has led to excessive lead intakes, especially when, in the past, lead toys and paints were more common, although this mode of ingestion can hardly be regarded as food. Nevertheless, Ullmann has regarded these sources as the major cause of lead poisoning among the children of Connecticut, even toothpaste tubes (but not the toothpaste) having been implicated. The total amount of lead added to the environment in the form of paint pigments and metallic products has been estimated as two to three times as much as is added in the form of the lead tetra-alkyl.[41] The fate of this lead is ill-defined, but it is probably mainly returned to the soil without passing through the atmospheric environment. An adult habit which can hardly be classed as *pica* is smoking: Patterson[5] has estimated that the average smoker will absorb 10 micrograms of lead per day more than a non-smoker.

INDIVIDUAL FOODS

Over 2000 individual samples of foods were examined for lead by local authorities in Britain in 1962–65,[42] including fruit, tomatoes, vegetables, and potatoes; in addition, over 200 samples of apples were examined in official laboratories.

None of the 429 samples of various foods examined by local authorities in England and Wales in 1966–68 contained lead in excess of the statutory limits.[43] Detailed results found for a wide range of individual foods in Bristol have been published[44] and show most samples to contain under 0·5 mg/kg. Pocklington and Tatton[45] examined apples imported into Britain from Australia, Canada, Italy, New Zealand, and South Africa during the period 1962–65 for lead residues, in the flesh, peel, and core. No sample was found to contain more than the statutory limit of 3 mg/kg lead in the peel or flesh, those in the flesh seldom exceeding one-half of this level.

Warren and Delavault[21] examined a variety of agricultural produce in both Britain and Canada and concluded that the oven-dried foods normally contained from 0·1 to 1·0 mg/kg lead, 2 to 20 mg/kg on the ash. In areas where, for natural reasons or by virtue of pollution, soils are abnormally high in lead, up to ten times these amounts may occur. In similar observations in the U.S.A., Kleinman[46] has made a statistical study of the variation of the lead levels found with the distance of the field or orchard from traffic, the traffic load, and the period of exposure. Average levels for a wide range of fruits, vegetables, cereals, and citrus crops seldom exceeded 0·2 mg/kg with no individual sample exceeding 1·0 mg/kg. Whilst the data suggested that the lead content of crops growing adjacent to traffic was greater than that of more distant crops, he rather cautiously concluded that other reasons for the presence of traces of lead needed to be taken more fully into account.

LEGAL POSITION IN BRITAIN

As indicated above, the lead content of food is controlled by the Lead in Food Regulations 1961[31] made under the Food and Drugs Act 1955. The lead content of paint on toys is controlled by the Toys (Safety) Regulations 1967[47] made under the Consumer Protection Act of 1961, being limited to a maximum of 0·5 per cent by weight. All of the pencils examined by the Greater London Council in 1970 were free from excessive lead and other toxic metals. The Department of Education and Science places a limitation on the lead content of pencils and crayons, which will also be controlled by regulations in due course. It has also been announced that the lead content of tinning on cooking utensils will be similarly controlled, a British Standard Specification[48] at present requiring the coating to contain not less than 99·75 per cent of tin.

CONCLUSION

In a recent report on airborne lead[41] the U.S. National Academy of Sciences concludes that, as for the waters of streams and lakes and for animal food products, there is no evidence that the amount of lead in the diets of people has changed substantially since 1940. Based on measurement of the lead levels in blood from 243 subjects from 19 locations within the U.S.A., Kubota et al.[49] concluded that the intake of lead from food and water was nearly the same by people who live in different cities. They found a mean blood level of 13·2 micrograms/100 ml: the exceptionally high values shown in the skew distribution were accounted for by exposure to sources of lead other than food and water.

REFERENCES

1. Lawther, P. J. *et al.* This conference.
2. Swaine, D. J. "The Trace-Element Content of Soils." Commonwealth Bur. Soil. Sci., Tech. Comm. No. 48, Harpenden, Herts, 1955.
3. "International Water Standards", World Health Organization, Geneva, 1963.
4. Ettinger, M. B. U.S. Public Health Symposium on Environmental Lead Contamination, 1965, pp. 22–27.
5. Patterson, C. C. *Arch. Envir. Health*, 1965, **11**, 344.
6. Annual Report of the Chief Medical Officer, Ministry of Health, 1967, p. 128, HMSO, London, 1968.
7. "European Standards for Drinking Water." World Health Organization, Geneva, 1970.
8. Packham, R. F. *Water Treat. & Exam.*, 1971, **20**, 144.
9. Murozumi, M., Chow, T. J., and Patterson, C. *Geochim. Geomochim. Acta*, 1969, **169**, 577.
10. Jaworowski, Z. *Nature, Lond.*, 1968, **217**, 152.
11. Mills, A. L. *Chem. in Britain*, 1971, **7**, 160.
12. Agricultural Research Council. "Effects of Air Pollution on Plants and Soils." HMSO, London, 1967.
13. Jones, J. S., and Hatch, H. B. *Soil Sci.*, 1945, **60**, 277.
14. Donovan, P. P., Feeley, D. T., and Canavan, P. P., *J. Sci. food Agric.*, 1967, **20**, 43.
15. Goodman, G. T. *Metals and Ecology Bull.*, No. 5, pp. 3–16, (1969), Swedish Natural Science Research Council, Stockholm.
16. Rains, D. W. *Nature, Lond.*, 1971, **233**, 210.
17. Mitchell, R. L., and Reith, J. W. S. *J. Sci. food Agric.*, 1966, **17**, 437.
18. MacLean, A. J., Halstead, R. L., and Finn, J. B. *Can. J. soil Sci.*, 1969, **49**, 327.
19. Motto, H. L., Daines, R. N., Chilko, D. M., and Motto, C. K. *Envir. Sci. Technol.*, 1970, **4**, 231.
20. Chow, T. J., *Nature, Lond.*, 1970, **225**, 295.
21. Warren, H. V., and Delavault, R. E. *J. Sci. food Agric.*, 1962, **13**, 96.
22. Bovay, E. *Mitteil. Lebens. Hyg.*, 1970, **61**, 303.
23. Goodman, G. T., and Roberts, J. M. *Nature, Lond.*, 1971, **231**, 287.
24. 1969 Report of the Government Chemist, p. 62. HMSO, London, 1970.
25. Hubbard, A. W., and Pocklington, W. D. *J. Assoc. publ. Anal.*, 1965, **3**, 29.
26. Kehoe, R. A. *Arch. Envir. Health*, 1961, **2**, 418.
27. Monier-Williams, G. W. "Lead in Food." Reports on Public Health and Medical Subjects No. 88. HMSO, London, 1938.
28. Egan, H., and Hubbard, A. W. *Chem. & Ind.*, 1971, 1181.
29. Food Chemicals Codex. National Academy of Sciences, Washington, D.C., 1966; Supplements 1967, 1968, 1969.
30. Report on Lead: Revised Recommendations for Limits for Lead Content of Food. Ministry of Food. HMSO, London, 1954.
31. The Lead in Food Regulations: Statutory Instrument 1961, No. 1931.
32. Oser, B. L. *Food cosmet. Toxicol.*, 1970, **9**, 245.
33. Bergner, K. G., and Miethke, H. *Z. Lebensm. Untersuch. Forsch.*, 1964, **125**, 406.
34. 1967 Report of the Government Chemist, p. 40. HMSO, London, 1968.
35. 1966 Report of the Government Chemist, p. 36. HMSO, London, 1967.
36. 10th Report, Joint FAO/WHO Expert Committee on Food Additives (1966), WHO Technical Report Series No. 373, Geneva, 1967.
37. "Cleaning Our Environment: The Chemical Basis for Action", p. 196. American Chemical Society, Washington, D.C., 1969.
38. Lewis, K. H. "Environmental Lead Contamination", p. 18. U.S. Public Health Service 1965.
39. Crawford, M. D., and Norris, J. N. *Lancet*, 1970, ii, 1087.
40. Ullmann, W. W. *Q. Bull. Assoc. food & drug Offic.*, 1971, **35**, 147.
41. "Airborne Lead in Perspective." U.S. National Academy of Sciences, Washington, D.C., 1971.
42. Advisory Committee on Pesticides and Other Toxic Chemicals: Report on the Collection of Residue Data. HMSO, London, 1969.

43 Association of Public Analysts. "Joint Survey of Pesticide Residues in Foodstuffs Sold
    in England & Wales." London, 1968, 1971.
44. City of Bristol Scientific Advisers Report, 1967, 1968.
45. Pocklington, W. D., and Tatton, J. O'G. *J. Sci. food Agric.*, 1966, **17**, 570.
46. Kleinman, A. *Pesticide Monit. J.*, 1968, **1**, 8.
47. Toys (Safety) Regulations 2967: Statutory Instrument 1967, No. 1157.
48. "Tin Coated Finish for Culinary Utensils." British Standard Specifications 3788:
    1964.
49. Kubota, J., Lazar, V. A., and Losee, F. *Arch. Envir. Health*, 1968, **16**, 788.
50. Monier-Williams, G. W. "Trace Elements in Food." Chapman & Hall, London, 1949,
    pp. 64–106.
51. Final Report of the Departmental Committee on Ethyl Petrol. HMSO, London, 1930.
52. Bagley, G. G., and Lock, L. N. *Bull. Envir. Contam. Toxicol.*, 1967, **2**, 297.
53. Merwin, B. W. *Ceramic Bull.*, 1971, **50**, 915.
54. Goldschmitt, V. M. "Geochemistry", *ed.* A. Muirs. Clarendon Press, Oxford, 1954.
55. Markland, J., and Vallance, J. *J. Assoc. Publ. Anal.*, 1971, **9**, 119.
56. Blokker, P. C. *Atmosph. Envir.*, 1972, **6**, 1.
57. Dedolph, R., Haar, G. T., Holzman, R., and Lucas, M. *Envir. Sci. Technol.*, 1970,
    4, 217.
58. Brandt, M., and Bentz, J. M., *Microchem. J.*, 1971, **16**, 113.

Discussion

A. J. Lambert (Burmah Oil Co.): I am rather interested to know the lead content of milk and what the situation is when it increases fourfold?

Dr H. Egan: I did in fact quote a figure from the Bristol Public Analyst, 0·06 per litre, but I am not quite sure of the significance of the second half of your question.

A. J. Lambert: Just really to total up the sources of lead from various foods and whether they are going to much exceed the 0·2 or 0·3 milligrams. Whether there is a connection between what one can assume from the calculations and the figures given, and the Part 2 and Part 3 figures.

Dr H. Egan: Indeed, the estimations of total daily intake of lead from food, and sometimes food and beverages, does in fact extend to beverages in each case, and of course is made up in just this way. To do this it is necessary to get figures, or best possible estimates on all the individual components of the daily diet and to consider them each in proportion to its average intake, and you can do this for all sorts of different populations too. But the figures which I have quoted seem to me to have been broadly stable over the last 30 to 40 years of, say, 300 micrograms a day for an adult, and are based on adding together the milk contribution as well as all the others, the milk contribution being respectively small.

A. J. Lambert: Another part of the point of asking the question on milk is that the cattle can take in 140 milligrams/day of lead, and yet the milk content is really quite low and they can even go to four times that 140.

Dr H. Egan: Certainly, under experimental conditions, cows fed on pasture which has been experimentally treated with lead, or taken deliberately from near highways where other measurements show it to be increased, have shown, as you say, up to four times the level in milk that they would have from other pastures.

Lead in the Environment: Possible Health Effects on Adults

By R. I. McCALLUM

(*Nuffield Department of Industrial Health, University of Newcastle upon Tyne*)

THE absorption of some inorganic lead by the human body has always been an inescapable feature of life on earth, although industrial activity in recent centuries has resulted in a general increase in exposure to lead by urban dwellers.[12] It is to be expected, therefore, that the body has developed defence mechanisms which can cope with a degree of lead absorption without serious harm but that some individuals will show intolerance to lead earlier or later than others. The situation is to some extent comparable with that of ionizing radiation, to which we have always been exposed, and which has, in the remarkably short period of about 75 years, increased as a result of technological change. In both cases there is a need to monitor and control the potential danger and to balance against this any benefits which accompany the risk. The difficulty with lead is to decide at what level of exposure absorption is such that a serious risk to health can occur in the most susceptible persons. The effects of lead on health have been studied most intensively in industry, both in the primary lead-producing factories and in a wide variety of secondary users. A brief consideration of industrial experience will serve to define the clinical problem in its most obvious form, and present the background against which current controversy is being carried on.

THE OCCUPATIONAL PROBLEM

Occupational exposure to lead is almost entirely an inhalation process in which dust or fume containing lead or its inorganic components is breathed and absorbed into the body through the lungs. Inhalation of organic lead compounds is a comparatively rare occurrence. In the past, occupational exposure to dust or fumes has been gross, so that clinical descriptions have depicted a range of gross symptoms and signs from severe lead colic, paralysis of hands or feet, and renal disease, to the cerebral form (encephalopathy) and death, as well as the less severe stippling of the red cells, anaemia, and a blue line on the gums.

The prevalence of occupational lead poisoning, as shown by the number of cases notified annually to the Chief Inspector of Factories, has markedly decreased since the 1920s. There has been a steep decline in notifications from a peak of 450–560 cases in 1924–25 to about 70 in recent years, and only one death has been ascribed to lead poisoning since 1950. There is no indication in the Annual Reports of the Chief Inspector of Factories of the severity of the great majority of notified cases but there is little doubt those being notified now are clinically much milder than those notified earlier in the century. This situation reflects a much improved occupational environment and much stricter criteria on which the diagnosis of poisoning is now made. As renal disease has disappeared and lead palsy become a rare condition, more attention has been paid to the detection of anaemia and chemical tests of blood and urine.

In the medical care of lead workers in industry we are now therefore concerned much more with laboratory tests than with clinical symptoms and signs. The aim in industry is the prevention of clinical poisoning by making sure that absorption of lead is kept low. The laboratory tests are to some extent a check on the environmental measurements of lead in air, but they are important also because the early clinical symptoms of lead poisoning are non-specific in character and difficult to evaluate. Furthermore, excessive occupational exposure to lead may give rise to a minimal amount of peripheral nerve fibre damage without clinical signs of nerve disease.[6]

LABORATORY TESTS

It is usual in lead workers to measure the level of lead in venous blood and in urine, either a spot sample or, better, a 24-hr sample, and to test quantitatively for coproporphyrins I and III or delta-amino laevulinic acid (ALA). These last two tests depend on the activity of the lead ion in interfering with the synthesis in the bone marrow of haem, the pigment which combines with protein to make haemoglobin. Excretion of ALA in the urine rises because lead inhibits the enzyme ALA dehydrase, which is a catalyst essential to the production of haem. Nakao and others[17] suggested that estimation of the enzyme should give an early indication of lead poisoning. The test, however, is very sensitive and does not correlate with symptoms of lead poisoning. It has not so far been used on any scale in the control of industrial lead absorption.

Haemoglobin estimation is a relatively crude biological test of lead exposure and is a useful screening procedure in many industrial situations but of no help at a low level of exposure.[27]

Measurement of coproporphyrins is perhaps the most frequently used biochemical test, usually in conjunction with a blood lead estimation. A quantitative test for coproporphyrins is quick and easy to carry out and quite sensitive. Estimation of ALA is reliable and rather more specific for lead. The results correlate well with metabolically active lead in the body and have a linear relationship with blood lead.[21]

In a statement on the diagnosis of inorganic lead poisoning[2], clinical findings supported by biochemical evidence of excessive lead absorption and evidence of exposure are given as the basis for a diagnosis. It is pointed out that many of the mild symptoms of lead poisoning are those of other complaints, some of them trivial. For example, tiredness, pallor, constipation, slight abdominal pain or discomfort, diarrhoea, poor appetite, sleeplessness, and irritability are symptoms experienced by very many people at one time or another and could arise from a variety of causes. Four categories of absorption are defined: normal, acceptable, excessive, and dangerous. Only the first two categories need concern us here (Fig 1). A blood lead of up to 40 micrograms per millilitre of whole blood is con-

TABLE I

Range of Normal and Acceptable Levels for Lead in Blood and Urine, and for Urine Coproporphyrin and Delta-amino Laevulinic Acid in Workers Exposed to Lead in their Jobs

Test	Normal	Acceptable
Blood lead	$< 40\ \mu g/100$ ml	$40–80\ \mu g/100$ ml
Urinary lead	$< 80\ \mu g/l$	$80–150\ \mu g/l$
Urinary coproporphyrin	$< 150\ \mu g/l$	$150–500\ \mu g/l$
Urinary δ-amino laevulinic acid	$0·6$ mg/100 ml	$0·6–2$ mg/100 ml

sidered normal, while 40 to 80 micrograms is thought to be an acceptable level for a lead worker. The corresponding figures for urinary lead, coproporphyrin, and ALA are also shown. The levels in the acceptable range apply to men who are in contact with lead in their jobs for eight hours a day and five days a week and in whom no clinical evidence of poisoning would be detectable. As they are absorbing lead, one may ask whether clinical criteria are sufficiently sensitive to enable us to continue to accept these levels and whether they are relevant to the problem of environmental lead and the general population.

THE THRESHOLD LIMIT VALUE FOR LEAD

The threshold limit value (TLV) for lead in air in the U.K. and U.S.A. is 0·2 mg per cubic metre and represents the time weighted concentration for an eight-hr day and a five-day working week. Although it is not a ceiling value which must never be exceeded, it is considered to be a borderline level for most people who work regularly with lead, and the stated level in some other countries is much lower. For example, in Russia the TLV is set at 0·01 mg per cubic metre and in Czechoslovakia at 0·05 mg per cubic metre.

Japanese investigators[23] have suggested a TLV of 0·12 mg per cubic metre for a 48- to 60-hr week or a little higher for a 40-hr week, with the aim of keeping urine coproporphyrin excretion below 50 $\mu g/l$. It would appear wise to regard the British TLV of 0·2 mg per cubic metre more as a safe ceiling level when considering levels of lead in air to which an urban population might be exposed.

LEAD ABSORPTION AND INTOXICATION

A clear distinction is often made between lead absorption and lead poisoning.[4] In lead absorption without any symptoms there may still be a mild anaemia, punctate basophilia, and reticulocytosis in the blood, an increase in lead content of blood, urine, and faeces, and of coproporphyrins in the urine. Gibson and others[8] refer to the difficulty in determining at what point a state of exposure to lead merges into a state of toxicity. Waldron[24] defines lead absorption as the uptake of lead by the subject from his environment by any route, and lead intoxication (pharmacological or clinical) as evidence that absorbed lead is interfering with some metabolic process in the body. This appears to make absorption identical with intoxication in the light of the evidence of interference with haem metabolism by very small amounts of absorbed lead. As it is no longer possible to maintain a clear division between absorption and poisoning, it might be better to abandon the effort. The use of biochemical tests has shifted the diagnostic emphasis in such a way that it might be preferable to talk of lead in terms of its chemical toxicity (without symptoms) and its clinical toxicity (with symptoms), the one merging into the other as a continuous process.

RISKS FOR THE GENERAL POPULATION

For some years there has been concern about the possible risk to the population at large from increased amounts of lead in air and food due to contamination of the environment from industrial processes, fuel combustion, and, in particular, the combustion products of organic lead compounds added to petrol. Much of the concern has been expressed in uncompromising terms by non-medical scientists, but a number of medical workers have been active in collecting much needed information on which to estimate the degree of risk. Although there are conflicting views among chemists on some of the findings relating to urban levels of lead in air, their source, and the degree of hazard associated with them, at least a factual collection of data on lead levels in air, food, and water is available about whose interpretation one can argue. On the other hand, the medical evidence on the lead hazard to the general population is inadequate.

In 1962, for example, there was a report from British Columbia that vegetation within a hundred yards of highways where the traffic was dense contained 100 times more lead than was normal. It was suggested that the lead could be ingested through vegetables, fruit, and cereals and that it might be responsible for complaints of dizziness, fatigue or tension, headaches, malaise, eyestrain, and defective vision.

A claim that the average resident of the U.S.A. is being subjected to "severe chronic lead insult"[18] introduced a new descriptive phrase and stirred up a considerable reaction from environmentalists. Patterson[18] defines a "typical" level of lead in blood as concentrations varying from "upper limits associated with acute intoxication", although no one is obviously ill, to "lower limits corresponding to levels which existed in man's prehistoric ancestors". "Natural" lead levels he describes as "equivalent to those which prevailed during the creation and evolution of our physiological response to lead". He introduces a metaphysical element into his argument by describing "contaminated" lead levels as those which have been "elevated above natural levels by man in activities which are an outgrowth of his abstract intellect".

Patterson claims that existing body burdens of lead (average 200 mg for a 70 kg adult) are 100 times larger than the natural burden, that the present rate of average lead absorption is 30 times the natural one, and that the source is atmospheric, mainly of industrial origin. He quotes a blood lead range in the U.S.A. of 0·05 to 0·4 ppm, with a mean of 0·25 ppm, and contrasts this with an average natural level of 0·002 ppm. The present-day mean of 0·25 ppm he thinks is much too close to what he calls the threshold for acute lead poisoning of 0·5 to 0·8 ppm of lead in blood. One may question his deduction of a natural level of lead in blood and at least insist that the natural range will be wide, depending on the variation in lead content of the soil in which food is grown. The way in which an average figure for blood lead in the general population is related to a blood lead level which is just within the acceptable upper limit for a lead worker is questionable.

Goldwater and Hoover,[9] in a study of lead levels in blood and urine from subjects in 16 different countries, gave a blood lead range from 15 to 40 micrograms with an average of 17 micrograms. Levels in urban subjects were generally slightly higher than in rural subjects. An investigation into lead concentrations in human tissue from 258 subjects who died in 33 different centres, ten in the U.S.A., the remainder in India, the Far East, and Africa, was reported by Schroeder and Tipton.[20] The samples were collected between 1952 and 1957. It was calculated that the median body burden of lead was 121 mg, but the amount tended to be higher in the U.S.A. than elsewhere. Most of the lead was found in bone, as would be expected. The amount varied very greatly from city to city, both in the U.S.A. and other countries. It accumulated with age in many of the soft tissues from American sources but not in tissues from other parts of the world, suggesting that in America absorption of lead between 1952 and 1957 was greater than the capacity of the body to maintain a steady lead balance. It was thought that in America the main cause of the increase was atmospheric lead from motor exhaust gases.

The concentration of lead in human tissues in the U.K. has been studied in 69 subjects at post-mortem.[1] The study was carried out in a densely-populated part of NW England, urban and industrialized and only four subjects had known occupational exposure. In adults the mean total body burden of lead was 162 mg in males and 113 mg in females. On average, nearly 95 per cent of this was in bone. In the soft tissues there was much less difference between the sexes, and the levels were relatively constant whatever the bone level. The authors comment that "the physiological capacity of human beings to control the absorption and retention of lead in the body would appear to be considerable. Man seems to be able to dispose of varying quantities of absorbed lead satisfactorily, without suffering ill effects".

LEAD IN AIR AND WATER

Patterson[18] gives a level of about 1·0 μg of lead per cubic metre in urban atmospheres and 0·05 μg of lead per cubic metre in rural ones, compared with a "natural" level of 0·005 μg per cubic metre. The amount of lead found in a London street by Waller et al[25] was 3·2 μg per cubic metre mean annual concentration, which was a level similar to those reported in cities elsewhere. They did not consider this an appreciable risk to health.

A recent estimate[3] for the level of lead in air for urban communities is up to 1·5 μg per cubic metre, of which 15 μg per day would be retained, but it would take a relatively large increase in atmospheric concentration to produce a small increase in blood lead.

About 30 μg a day is contributed by food and water, and people who live in soft water areas have more lead in their bones than those from hard water areas. Concern has been expressed about the acidity of water in the British Isles, because much of it can have a pH of less than 7·6 and could dissolve lead,[19] but water supplies in Britain do not apparently contain more than the international standard of 0·05 ppm of lead.

Thompson[22] measured the quantity of lead ingested by five people not exposed to lead at work or by accident and found a wide variation from day to day of 70 μg to 750 μg (average 274 μg) in the amount of lead in food. About 90 per cent of the total intake of lead was excreted in the urine or faeces but an average of about 10 μg was retained.

LEAD FROM PETROL

Much of the concern about lead exposure of the general population comes from non-medical scientists and journalists who consider lead from petrol the chief source. Bryce-Smith,[5] for example, criticizes the petroleum industry for their attitude to the levels of lead found in the general population. He underlines the wide range of individual reaction to a poison such as lead which makes any estimate of an acceptable level of absorption of limited value. Bryce-Smith puts the blame for a rising proportion of lead in city dust mainly on the use of leaded petrol but at the same time admits that it is not the sole source.

He also questions the safety of tetraethyl lead on the grounds that there is a lack of information on "the specific psychic symptoms which might have been expected". He goes on to claim that no other toxic chemical has accumulated in humans to average levels so close to the point of potential clinical poisoning, and that some symptoms in the general population are already attributable to lead. These symptoms he describes as depression, headaches, and undue fatigue, and claims that they have been cured or improved in 85 per cent of cases by treatment with a chelating agent for lead which removes it from the body.

In contrast, Mills[16] considers that the potential risk to the urban population from leaded petrol is negligible. Inorganic lead salts in the exhaust gas of cars vary in quantity according to engine speed, tending to be retained at low speeds. After emission the lead in air falls rapidly, and car exhaust contributes an insignificant quantity of lead to drinking water. Mills accepts the view that levels of lead in the blood of the general population have not risen for about 40 years and that these levels are within an acceptable range.

While agreeing with this, one must take into account that it is unsafe to rely solely on blood levels because they represent only one phase of a dynamic situation in the body in which lead is passing from the lungs or gut into the blood, from the blood to the tissues and back, and to bone, where it is stored but is still in a dynamic equilibrium with the blood; at the same time, lead is being excreted from the body, mainly through the kidneys. For this reason, at least, one other of the indices of lead activity in the body (Fig 1) needs to be considered as well.

The source of airborne lead has been investigated by a technique using relative proportions of lead isotopes.[7] The average isotopic ratios for petrol correlated well with air or soil in ten different countries in Europe, America, and Asia. Tinker,[7] in an article which pre-judges the issue, thinks that a correlation between the isotopic ratios of lead in petrol and lead in human tissues will be found but he admits that this investigation has not yet been carried out. Chow's work has been criticized for containing serious inconsistencies[13] and its confirmation by other workers appear necessary.

Williams,[26] in a reply to Tinker's article, quotes a mean level of lead in air in a city street of 3 to 5 micrograms per cubic metre or about a fortieth of the TLV. He argues that although absorbed lead is cumulative, this makes it less hazardous by allowing time for its detection and measurement and for remedial action, that lead poisoning in children is nearly all due to ingestion rather than inhalation, and that the risk from present blood lead levels has been exaggerated; and he questions the conclusions drawn by Bryce-Smith that raised blood lead levels in American cities are due mainly to lead from petrol. Finally, he criticizes the view that an increase in mental illness is related to a rise in the level of lead in air on the grounds that there is little evidence that either the lead contamination is increasing or mental illness (as opposed to admission rates to mental hospitals) is increasing.

ALA DEHYDROGENASE

A new factor to be considered is the application of the ALA dehydrogenase test to people not occupationally exposed to lead.

Hernberg and Nikkanen[11] described measurable changes in the erythrocyte ALA dehydrogenase in 26 healthy medical students who had never had any occupational exposure to lead but only to ordinary environmental pollution. This pollution is described as "normal urban conditions", but the actual level of lead in air is not stated. The amount of ALA dehydrogenase in red blood cells is shown to fall as the blood lead level rises and that the fall is proportional to the concentration of lead in blood. This test may well have a valuable place in biological monitoring of general urban pollution from lead by using erythrocyte ALA dehydrogenase measurements in a sample of the human population, together with similar studies in animals.[8]

The effect of lead on this enzyme in the brain might be of much greater importance than interference with its activity in red cells,[14] but it cannot be studied in human brain.

CONCLUSIONS

While there is considerable argument over the amount, source, and significance for health of lead in the air in cities, and in the bodies of their citizens, there is at least agreement on the need for more data on them. Regular monitoring of these indices and of the output of airborne lead from different sources is a reasonable requirement. The point has been well made that a continued increase in the level of environmental lead could at some future time result in some people not occupationally exposed to lead coming into the range of acceptable levels of lead exposure for industrially-exposed workers. This is far removed from clinical poisoning as at present conceived. It is necessary, however, to consider whether more refined clinical tests would show some undesirable lead effect at an earlier stage than is possible at present.

No sound medical evidence has been produced so far to support the prevalence of any significant amount of clinical or sub-clinical lead poisoning in the general population. Emotive, uncritical, and exaggerated claims made on little or no evidence do not help. The suggestion that an illness of the general population characterized by depression, headache, and fatigue is caused by the present levels of urban atmospheric lead is almost incapable of proof or disproof because symptoms of this kind are notoriously difficult to assess. The view that ". . . the normal state of most people is to feel greatly tired, harassed, and under the weather . . ."[15] sums up this problem neatly.

However, early symptoms of lead poisoning are so indefinite that it might be worth while to investigate the use of a battery of psychological tests in men exposed to a known lead hazard but without clinical poisoning. Such tests have been found useful in the detection of latent carbon disulphide poisoning,[10] which presents as a group of mental changes such as depression, mental retardation, intellectual impairment, and mild motor disturbances. The tests could be related to blood lead levels, urine coproporphyrin, and ALA excretion and might provide a much more sensitive index of clinical abnormality from either occupational or general urban lead exposure than is at present available.

REFERENCES

1. Barry, P. S. I., and Mossman, D. B. *Brit. J. Industr. Med.*, 1970, **27**, 339–51.
2. "Diagnosis of Inorganic Lead Poisoning: a Statement". *Brit. med. J.*, 1968, **2**, 501.
3. "Lead in Air". *Brit. med. J.*, 1971, **2**, 653–54.
4. Browning, E. "Toxicity of Industrial Metals". Butterworth, London, 1961.
5. Bryce-Smith, D. *Chem. in Brit.*, 1971, **7**, 284–86.
6. Catton, M. J., Harrison, M. J. G., Fullerton, P. M., and Kazantzis, G. *Brit. med. J.*, 1970, **1**, 80–82.
7. Chow, T. J. Reported in *New Scientist* and *Science Journal*, by Tinker, J. 1971, **51**, 323.
8. Gibson, S. L. M., MacKenzie, J. C., and Goldberg, A. *Brit. J. industr. Med.*, 1968, **25**, 40–51.
9. Goldwater, L. J., and Hoover, A. W. *Arch. environ. Health*, 1967, **15**, 60–63.
10. Hänninen, H. *Brit. J. industr. Med.*, 1971, **28**, 374–81.
11. Hernberg, S., and Nikkanen, J. *Lancet*, 1970, **1**, 63–64.
12. Kehoe, R. A. *J. Roy. Met. Pub. Hlth*, 1961, **24**, 81–96, 101–20, 129–43, 177–203.
13. Kirton, H. M. *New Scientist & Sci. J.*, **51**, 772.
14. "Lead in Air". *Lancet*, 1971, **2**, 653–54.
15. Miller, H., in "Medical History and Medical Care", *ed.* G. McLachlan and T. McKeown. London, 1971, p. 224.
16. Mills, A. L. *Chem. in Brit.*, 1971, **7**, 160–62.
17. Nakao, K., Wada, O., and Yano, Y. *Clin. Chim. Acta*, 1968, **19**, 319–25.
18. Patterson, C. C. *Arch. environ. Health*, 1965, **11**, 344–60.
19. Reed, C. D., and Tolley, J. A. *Lancet*, 1967, **1**, 894.
20. Schroeder, H. A., and Tipton, I. H. *Arch. environ. Health*, 1968, **17**, 965–78.
21. Selander, S., and Cramér, K. *Brit. J. industr. Med.*, 1970, **27**, 28–39.
22. Thompson, J. A. *Brit. J. industr. Med.*, 1971, **28**, 189–94.
23. Tsuchiya, K., and Harashima, S. *Brit. J. industr. Med.*, 1965, **22**, 181–86.
24. Waldron, H. A. *Brit. J. industr. Med.*, 1971, **28**, 195–99.
25. Waller, R. E., Commins, B. T., and Lawther, P. J. *Brit. J. industr. Med.*, 1965, **22**, 128–38.
26. Williams, M. K. *New Scientist & Sci. J.*, 1971, **51**, 578–80.
27. Williams, M. K., King, E., and Walford, J. *Brit. med. J.*, 1968, **1**, 618–21.

Discussion

Professor D. Bryce-Smith (University of Reading): The speaker referred sarcastically to the absence of any claims that lead caused impotence. Is he aware of Russian work on this matter, *e.g.* by Egorova and Galubovich,* who demonstrated testicular and sperm damage in experimental animals at an extremely low dosage of lead, and that the incidence of impotence in industrially exposed males has been found to be related to the severity of exposure to lead?

Dr R. I. McCallum: I am not familiar with the Russian work which has been quoted. In referring to impotence, I pointed out that it had not been mentioned in the lists of symptoms in the general population which some authors (*e.g.* Bryce-Smith[5]) quoted as being due to lead. This absence is surprising because impotence is a common problem from purely internal causes, and is often ascribed irrationally to external factors.

Dr A. Berlin (Commission of the European Communities): In his introduction Dr McCallum seems to suggest that the body has developed defence mechanisms to cope

* Egorova, G. M. *et al.*, *Toksikologiya novykh promyshlennykh khimicheskikh veshchestv.*, 1966, **8**, 33; Galubovich, E. Ya. *Ibid.*, 1968, **10**, 64.

with absorbed lead in the same way that it has developed defence mechanisms against ionizing radiation.

Not being aware of such mechanisms in the case of ionizing radiation, it would be useful if Dr McCallum could clarify this point.

Dr R. I. McCallum: I agree that my statement needs some clarification. What I wished to convey was that the human race has been able to adapt itself to a wide range of quantities of absorbed lead over a very long period and has survived, just as it has been able to adapt to certain levels of ionizing radiation. What has happened in recent years is that the size of the dose has increased and the point at which it must not be allowed to increase further has to be determined.

Dr N. McNeil (Scottish Home and Health Dept, Edinburgh): Dr McCallum states that "water supplies in Britain do not apparently contain more than the international standard of 0·05 ppm".

In fact, in Scotland, several cases of clinical and sub-clinical lead poisoning have been found in rural areas of Scotland and also the city of Glasgow, where the water supplies are soft, and where lead service pipes and lead storage tanks are still in use in some older properties.

The cases all exhibited depression of the ALA dehydrase in the blood.

Dr R. I. McCallum: Even if water as supplied to the house is safe, then lead may be added subsequently from the domestic system. This type of poisoning incident therefore occurs sporadically and the remedy is clear.

At present the significance of the red cell ALA-dehydrase test in people without clinical evidence of lead poisoning is not clear.

Children and Environmental Lead

By DONALD BARLTROP

(Paediatric Unit, St Mary's Hospital Medical School, London, W.2)

PREVIOUS work on the potential hazards of exposure to inorganic lead has been confined to observations on adult male workers in the lead industries and some limited studies on adult males acting as volunteers in laboratory situations.[1,2] Results obtained in this way are not capable of general application, since the population is non-homogeneous and includes groups which are obviously distinct from the typical employee in the lead industries in respect of their interaction with lead. Thus the infirm, the elderly, the malnourished may all react differently from the healthy adult male worker on exposure to lead. Adult women have long been excluded from the lead industry because they were thought to be more susceptible to lead posioning than their male counterparts. This paper, however, is concerned with another and quite distinct population group—the immature, in which should be included the entire non-adult population ranging from fetus to the child of school age. The justification for examining this group lies in the fact that it is the only group in this country in which death and cerebral damage resulting from lead poisoning continue to occur. It is thus important to examine the child's environment and his interaction with it in relation to growth and development.

PHYSIOLOGICAL ASPECTS

The quality that best characterizes childhood is that of growth. The child is constantly changing his stature at a rate which varies with age and in addition the relative proportions of the child's body to each other also undergoes continuous change. Changes in stature and shape are of special relevance to the skeletal tissues which are being re-modelled and re-fashioned throughout development. It is of interest to recall that 90 per cent of the body burden of lead is deposited in the skeleton and must be involved in some of these changes, although it remains unknown, whether or not significant amounts of lead are actually released from the skeleton during growth.

Non-skeletal tissues are also involved in growth and remarkable differences between their relative growth rates occur in comparison with each other and with general somatic growth. Thus neural tissue has almost completed its growth by the end of the second year, whereas genital tissues are relatively retarded until shortly before puberty. Lymphoid tissue actually exceeds its adult size during childhood and regresses after the seventh year.[3] Superimposed on these gross structural changes are the less readily discerned changes in cell chemistry involved that are associated with biochemical maturation. Thus certain drugs cannot be given to the new-born because the detoxication mechanisms of the liver are not functional, although they may be given safely two to three weeks later.[4]

It follows that the effect of an extrinsic toxin such as lead might well involve different tissues at different stages of development and affect a particular tissue in different ways according to the maturation of the constituent cells.

The exposure of children to lead in the atmosphere or diet may result in different retentions of lead than similarly exposed adults and this again is a reflection of the physiology of childhood. It is a commonplace observation that the dietary needs of children may exceed those of adults and this may be related to the high energy output of most children and to the additional requirements of growth. It follows that on a body weight basis the child will ingest more lead than an adult eating the same diet. A similar deduction has been made with respect to atmospheric lead.[5] Verification for the dietary figures is available as a result of studies in which the faecal excretion of lead by children has been measured, since 90 per cent of ingested lead is excreted unabsorbed. Mean faecal lead contents of 132 μg/24 hr[6] and 130 μg/24 hr[7] have been reported in children not known to be abnormally exposed to lead, the latter study recorded the range of 123–183 μg lead/24 hr for children aged two to three years. Comparable studies for airborne lead in childhood are not available and while there is general agreement concerning the intake of lead experienced from air and water by some age groups, there is no agreement concerning absorption of the lead intake. The greatest controversy surrounds airborne lead and estimates for absorption of suspended lead particles from the respiratory tract have ranged from 10–50 per cent in adults. However, there is no means of relating this to the anatomical differences in the respiratory tracts of children as opposed to adults. It is not yet possible, therefore, to relate particular rates of absorption of lead to intake in children, nor is it possible to estimate the blood lead concentration corresponding to particular environmental conditions with any degree of precision.

Special physiological problems surround the most immature and rapidly-growing members of our species—the fetus. Dietary and respiratory lead cannot directly affect the fetus but do so through the intermediary of the maternal circulation and the placenta. Although the placenta has some selectivity for the minerals that it transmits, this does not seem to apply to lead, which has been shown to traverse the placental barrier in the rat[8] and in the human.[9] Exposure to environmental lead occurs from very early stages of fetal development, certainly by the sixteenth week, although presumably the fetal tissues receive only a share of the total lead absorbed by the mother.

BEHAVIOURAL FACTORS

The interaction of an individual with his environment must depend in part upon the nature of his activities within it. Children, as with the young of other species, are generally encouraged to explore their surroundings, but the manner of this exploration will vary with the age of the child and the character of his environment. The toddler that investigates every accessible part of his surroundings to the delight or embarrassment of his parents is as familiar as are the incessant intellectual forays of the inquisitive pre-school child. These activities combine both enquiry and movement to a varying degree and may occur independently or in "play", which is a far more important activity than is generally realized.

Very young children are limited in the scope of their activities by their lack of physical strength, co-ordination, and mobility. However, they utilize an additional means of exploration, not wholly abandoned in adult life, the mouth. Observation of a young child reveals that the mouth is used as a device for sensing the shape and texture of materials as well as their taste. The activity of "mouthing" should be distinguished from that of swallowing or ingesting the materials sampled in this way, although this may not be valid in connection with objects coated with a lead-containing paint. The practice of ingesting materials not normally regarded as food is known as pica, a reference to the magpie (pica pica pica) which is an inquisitive and adventurous eater. This activity is not confined to children, in whom it may be regarded as a normal part of development, but it may persist into adult life in certain racial groups, in pregnancy, and in extreme hunger.[10]

Pica may differ in quality and degree from one child to the next and parents will confirm the remarkable list of objects that they have observed their children or ingest during a 24-hr period, including paper, household dust, coal, soap, cigarette butts, soil from the garden, or household plants.[11] Similarly, approximately 50 per cent of children aged two to three years in London had been observed to have oral contact with a painted surface during a similar period.[12]

Although pica probably represents a normal phase of maturation in the majority of situations analogous to thumb-sucking, other associations may exist. It has been suggested that the practice represents an attempt to replace deficient minerals such as iron or calcium and there is certainly evidence that mineral-deficient animals will "self-select" an optimal diet to satisfy their mineral needs.[13] This is most plausible in adults who consume chalk or particular soils (geophagia)[14] but is difficult to reconcile with the childhood situation. Some evidence has been presented that children with pica are iron-deficient,[15] come from disturbed families, or particular racial groups, but the prevalence of the practice would not support most of these suggestions.[11] Mental retardation itself may be associated with pica of unusual persistence or duration, so that such children have been termed "scavengers". The most bizarre and extreme forms of this self-destructive gnawing of the lower lip or limbs occurs sometimes, but not always, in association with an inborn error or uric acid metabolism (Leisch-Nyhan syndrome). The blood lead concentration of children with pica reflects the lead content of the materials ingested so that some of these children have higher blood lead concentrations than those in control groups. Although some workers suggested that the increased blood lead concentration in some children was the cause of their mental retardation,[16] other studies have failed to show significant differences between retarded children and controls.[17]

It is not appropriate to make a more detailed analysis of pica in this paper but it should be recognized as a common activity in childhood that results in a different relationship with the environment for those that indulge in it compared with those that do not. Since all children pass through a phase of oral exploration involving pica at some stage in their development, this may be relevant to their lead intake during early life.

THE CHILD'S ENVIRONMENT

The term "environment" has come to acquire a narrow connotation in connection with the contamination of air, soil, and water by industrial and domestic wastes. This has diverted attention from the concept of "environment" in the natural history sense, which has been defined as the "sum of external influences affecting the organism". The domestic environment, in terms of surface coating materials in the home, soil chemistry in the garden, and the nature of toys and playthings, is as important for children as the wider environment shared with adults.

The most important source of lead for children in both content and accessibility is lead paint in the home applied to surfaces, such as window sills, doors, and furniture. Typically, this problem is greatest where old paint of high lead content has been allowed to flake and peel, so that children inhabiting old, poorly-maintained properties have ready access to this source. Defective plaster work coated or impregnated with lead paint is another hazardous material. In some cities in the U.S.A., whole districts of old, sub-standard housing are recognized as "lead-belts", so that the affected children are often derived from the under-privileged section of the community.[18] A survey in part of London, however, demonstrated, in that district at least, that lead paint is not confined to old housing or to the homes of poor families. Some 53 per cent of samples contained more than 1 per cent lead and 25 per cent contained more than 5 per cent lead. Appreciable numbers of the samples were derived from relatively modern housing and from homes of high socio-economic status.[20]

TABLE I

Comparative Measurements for Child and Adult (after King, 1971)[5]

Age	2 yr	Adult	%
Weight, kg	12·6	70	18
Calories/24 hr	1280	2700	47
Estimated lead intake, μg/24 hr			
diet (a)	150	300	50
air (b)	6	15	40

(a) Assuming lead intake to be 50 per cent of adult diet.
(b) Assuming resting ventilation of air containing 2 μg/m³/24 hr.

Much of the lead paint problem is connected with the incomplete removal of old paint films at redecoration, so that re-painting may actually preserve the old deep layers of paint of high lead content *in situ*. The older the house, the more likely it is to contain paint of high lead content, so that in the district studied the highest lead contents were found in homes built prior to 1855. Re-painting with lead paint has an additional result in that as the paint film increases in thickness the area of the film containing unit mass of lead decreases and a flake of paint of only a few millimetres diameter may contain more than one milligram of lead. Comparative figures may be derived for a two-year-old child (Table I) and it can be calculated that an equivalent lead intake would be contained in the total diet for seven days combined or in the total air volume respired in 170 days.

Toys and furniture are important features of the domestic environment for children and may be painted. Regulations are in force limiting the lead content of paint applied to toys and nursery furniture for sale to 0·5 per cent in the dried film.[19] This limit would appear, however, to have been arrived at with little regard to toxicological criteria. It is apparently still possible to import toys into the U.K. which are decorated with lead paint. Paints of high lead content are still manufactured and sold in the U.K. in receptacles that bear a warning label. The advice of the manufacturer concerning the application of such paints is not always heeded, so that some paints of high lead content inevitably find their way into the home on surfaces accessible to children. It is perhaps pertinent to add that children are not the only group exposed to this hazard but that domestic animals including dogs[20] and calves[21] are also liable to lead poisoning from paint.

The increasing concern for this environmental source of lead has resulted in the search for simple, non-destructive techniques for the determination of lead in paint films *in situ*. It is preferable to identify hazardous environments before children inhabiting them have acquired an appreciable body burden of lead. Screening programmes based on blood lead analysis must always be unsatisfactory since, in effect, they utilize children to identify homes with sub-standard paintwork. One approach has been the application of a device well known to geologists, the portable isotope fluorescence analyser. This device contains a small isotope source which excites a characteristic X-radiation from the lead in the surface to which it is applied. Using a plutonium source of 30 mCi, it is possible to obtain remarkably accurate results for simple paint films. However, less satisfactory results are obtained in the field where film non-homogeneity is common and the deepest part of the film is screened from the exciting source by the superficial layers.[22] It is likely, however, that further developments of this type of apparatus will increase its effectiveness.

The soil has, until recently, been neglected as a significant source of lead for children. It arises in a variety of ways but local lead-containing ores and waste materials from mining and smelting operations are well-recognized sources of contamination. In some circumstances in North Wales and in the Pennines the lead mining and refining activities took place during the last century but active smelting operations in South Wales and in London have recently been shown to be contributing lead to the local environment. Coal burning and leaded petrol also contribute to the lead content of soil that may ultimately be used for housing. Soils in the U.K. seldom contain more than 200 ppm lead but there are local variations, so that areas with soil leads in excess of 3000 ppm are not uncommon. Recently, some soils in urban situations have been shown to contain lead in concentrations of up to 20,000 ppm (2 per cent). Children in such districts are at risk largely through the possibility of ingestion either as a result of pica for garden soil or inadvertently by oral contact with hands or playthings contaminated with soil or dust. A case of lead poisoning from garden soil has been reported in London[23] as a result of pica, but at present the alternative possibility remains speculative. In common with adults, the ingestion of soil on imperfectly prepared vegetables grown in such soil or, in some districts, contamination of the local water supply are possible. If soil containing 20,000 ppm lead is compared with that available in lead paint, then it can be calculated that 50 mg of soil (1 mg lead) daily will result in lead poisoning after five to six months[7] and that lesser amounts of the order of 25 mg daily (500 μg lead) would be sufficient to increase the blood lead concentration of young children signficantly.[6]

Although airborne lead in most situations makes only a modest contribution to the lead intake of children, there are circumstances in which significant intakes can occur. These are usually in relation to local situations in which lead or lead-contaminated material is volatilized within or in proximity to children's homes. A rare but recurrent source of domestic lead poisoning source is the burning of battery cases as fuel in times of economic distress[24]. Children would seem to be liable to severe and even fatal poisoning in this situation, although the whole family is exposed, suggesting that either they experience a more prolonged exposure unrelieved by leaving home for work during the daytime or that there is some additional factor such as lead-contaminated dust arising from the ashes. Exposure to lead fumes has been reported in other circumstances, e.g. children actually living in a family print shop[25] and the domestic manufacture of lead weights for fishing tackle.[26] Older children undergoing vocational training in the assembly of electronic components have absorbed lead from solder[27] and young girls employed in the manufacture of jewellery have been poisoned from the lead frit employed in enamelling.[28] The contribution of motor vehicle exhaust fumes to the inhalational lead intake of children has yet to be determined, but it would appear to be insignificant with regard to other sources of lead in the child's environment.

Among the more bizarre sources of lead affecting children have been white lead dusting powders,[29] lead nipple shields employed by some Italian nursing mothers,[30] and the black "eye-liner" that adorns very young children from the Indian sub-continent, some specimens of which contained almost pure lead sulphide.[31] In rural districts of Yugoslavia, children have been poisoned from wine stored in lead-glazed vessels.[32]

The significance of various environmental sources of lead for children is difficult to determine and there are obviously regional variations dependent on geographical location and local custom. Experimental studies, such as those of Kehoe with adult volunteers,[2] have not been repeated in children, so that the proportion of the lead intake actually absorbed from the gut or respiratory tract is not known with certainty. It is generally assumed in adults that at least 90 per cent of ingested lead is excreted unabsorbed (10 per cent retained) but estimates for the retention of atmospheric lead have varied from 10 to 50 per cent. Applying these figures to the data in Table I and assuming 50 per cent absorption of

TABLE II

Absorption of Lead by Children aged 2 Years

Source	Intake μg/24 hr	Absorption μg/24 hr
Diet	150	15
Air (2 μg/m^3)	6	3

atmospheric lead, it is possible to calculate the relative contribution of these two sources in children not otherwise exposed to lead (Table II). To put these figures in perspective, a survey of the atmospheric content of lead in three cities in the U.S.A. showed a mean range of 1–3 μg/m^3 during the period 1961–62,[38] so that even assuming the maximal possible absorption of airborne lead, it would require a fivefold increase in the airborne lead concentration before absorption equalled that derived from the normal diet.

TISSUE CONCENTRATION AND TOXICITY

Lead absorbed from the gut is initially bound to certain soft tissues including the erythrocytes, liver, and the kidney. Ultimately, much of this lead is transferred to bone, where it is metabolically inactive, and the remainder is excreted in the urine and bile. Although the lead content of a variety of tissues have been determined, including hair[33] and deciduous teeth,[34] the blood lead is probably the best index of soft tissue concentration. Biochemical changes can now be demonstrated at very low blood lead concentrations, *e.g.* the inhibition of ALA dehydratase activity,[35] but the significance of this is uncertain. The situation has also been complicated by the rapid changes in analytical techniques in recent years which now permit the instantaneous determination of lead in a few μl of blood, so that results obtained in different centres at different times are not always comparable. Even the "normal" blood lead for children is subject to controversy, since this has yet to be determined in a group of children in whom the dietary and atmospheric sources of lead have been controlled and in whom the variables of age and season have been recognized. Most available evidence suggests that children do not differ from adults in that symptomatic lead poisoning is unlikely at blood lead concentration below 80 μg/100 g and encephalopathy unlikely at concentration below 100 μg/100 g. However, it is probable that 5 per cent of children or more in some urban communities have blood lead concentrations in excess of the usually accepted "normal" limit of 40 μg/100 g but that most of them are asymptomatic. There is as yet no evidence that long-term elevation of the blood lead concentration within the 40–80 μg/100 g range has any biological significance, although such children would require a much smaller additional daily burden of lead to precipitate frank poisoning than those with lesser values. There is no doubt, however, that those who develop a lead encephalopathy, and survive, may sustain permanent cerebral damage.[25] There is also evidence that chronic nephritis occurs more frequently in those who experienced symptomatic poisoning during childhood.[36]

The frequency with which encephalopathy occurred in children as opposed to adults has given rise to speculation that children are more susceptible to lead intoxication than adults. This view has been fortified by the severity of lead poisoning in children in family exposures to lead and the death of the fetus after maternal exposure.[37] This discrepancy with regard to children probably reflects the improved health control measures that have been implemented in the lead industries during the last 50 years, together with the inherent problems in paediatric diagnosis. The lead poisoning of adult "moonshiners" who consume illicit whisky prepared in lead-contaminated stills has much in common with the paediatric disease including encephalopathy.[38]

The fetus is known to absorb lead by placental transfer from the mother,[9] although the significance of this for the human is not known. Developmental abnormalities corresponding to spina bifida in the human have been demonstrated in the hamster after exposure to lead,[39] but no comparable relationships have yet been detected in human populations.

CONCLUSION

Children constitute a distinct population group which is characterized by certain physiological and behavioural characteristics. Considerations of the environmental sources of lead for children should include the home and domestic surroundings, in addition to environmental factors relevant for adults. Lead paint frequently, and soil lead occasionally, may make substantial additions to the daily lead intake in children, although many less common additional sources have been described. The relative contributions of atmospheric and dietary lead to the soft tissue lead concentration of children has not yet been established. Although children have not been shown to be more susceptible to lead than adults, the situation concerning the fetus deserves further study.

REFERENCES

1. Lane, R. E. *Arch. environ. Hlth*, 1964, **8**, 243.
2. Kehoe, R. A. *J. roy. Inst. Publ. Hlth*, 1961, **24**, 177.
3. Scammon, —. "The Measurement of the Body in Childhood". The Measurement of Man, University of Minnesota Press.
4. Shirkey, H. C., Vaughan, V. C., and McKay, R. J., in "Textbook of Pediatrics", *ed.* W. E. Nelson. Saunders, Philadelphia, 1969.
5. King, B. G. *Amer. J. Dis. Childh.*, 1971, **122**, 337.
6. Chisolm, J. J. Jr. *Pediatrics*, 1956, **18**, 943.
7. Barltrop, D., and Killala, N. J. P. *Lancet*, 1967, **2**, 1017.
8. Baumann, A. *Arch. Gynaek.*, 1933, **153**, 584.
9. Barltrop, D., *in* "Mineral Metabolism in Paediatrics". Blackwell, Oxford, 1969, pp. 135–54.
10. Cooper, M. "Pica". Charles C. Thomas, Springfield, Illinois, 1957.
11. Barltrop, D. *Amer. J. Dis. Childh.*, 1966, **112**, 116.
12. Barltrop, D., and Killala, N. J. P. *Lancet*, 1967, **2**, 1017.
13. Richter, C. P. *Harvey Lect.*, 7942–1943, **38**, 63.
14. Ferguson J. H. and Keaton, A. G. *New Orleans Med. Surg. J.*, 1950, **102**, 460.
15. Lanzkowsky, P. *Arch. Dis. Childh.*, 1959, **34**, 140.
16. Moncrieff, A. A., Koumides, O. P., Clayton, B. E., Patrick, A. D., Renwick, A. G. C., and Roberts, C. E. *Arch. Dis. Childh.*, 1964, **39**, 1.
17. Gordon, N., King, W., and MacKay, R. I. *Brit. Med. J.*, 1967, **1**, 480.
18. Griggs, R. C., Sunshine, I., Newill, V. A., Newton, B. W., Buchanan, S., and Rasch, C. A. *J.A.M.A.*, 1964, **187**, 703.
19. Toys (Safety) Regulations, 1967.
20. Berry, A. P. *Vet. Rec.*, 1966, **79**, 248.
21. Ministry of Agriculture, Fisheries and Food, 1959, Animal Health leaflet, No. 43.
22. Barltrop, D., Harford, C. L., and Killala, N. J. P. *Bull. Environ. Cont. & Toxicol.*, 1971, **6**, 502.
23. Orton, W. T. *Med. Officer*, 1970, **123**, 147.
24. Travers, E., Rendel-Short, J., and Harvey, C. C. *Lancet*, 1956, **2**, 113.
25. Byers, R. K. *Pediatrics*, 159, **23**, 585.
26. Heese, B. Personal communication, 1971.
27. Lyubchenko, P. N. *Gig. Sanit.*, 1966, **31**, 93.
28. Fothergill, R., Kipling, M. D., and Weber, A. B. *Brit. J. ind. Med.*, 1967, **24**, 333.
29. Holt, L. E. *Amer. J. Dis. Childh.*, 1923, **25**, 229.
30. Ammaniti, L. and Longobardi, G. *Arch. Ital. Pediat.*, 1962, **22**, 241.
31. Roots, L. M. *Med. Officer*, 1969, **122**, 87.
32. Beritic, T. Personal communication.
33. Kopito, L., Briley, A. M., and Shwachman, H. *J.A.M.A.*, 1969, **209**, 243.
34. Strehlow, C. D. Proc. 15th Ann. Bioassay Meeting, Los Alamos, LA-4271-MS, 1969
35. Hernberg, S., and Nikkanen, J. *Lancet*, 1970, **1**, 63.
36. Henderson, D. A. *Aust. Ann. Med.*, 1954, **3**, 219.
37. Wilson, A. T. *Scott. med. J.*, 1966, **11**, 73.
38. U.S. National Academy of Sciences. "Airborne Lead in Perspective". Report by the National Research Council, 1971.
39. Ferm, V. H., and Carpenter, S. J. *Exp. molec. Pathol.*, 1967, **7**, 208.

Discussion

Professor D. Bryce-Smith (University of Reading): Dr Barltrop has remarked that there is no evidence to indicate whether children are more sensitive than adults to lead. Is he aware of the work of Kehoe,* who give adult male volunteers 2 mg of lead per day for several years without, as he claimed, any ill-effects? This compares with the intake of 1 mg/day which Dr Barltrop finds will produce clinical poisoning in a child within a few months.

Dr D. Barltrop: Kehoe's work was based on a limited number of long-term balance studies in adults during the 1930s. Analytical techniques have changed greatly since then and perhaps this type of study could, with advantage, be repeated. A comparable lead intake for an adult (on a surface area basis and assuming a dietary intake of 200 μg lead/day in the child) would in fact be about 4·1 mg/day.

Dr G. Kazantzis (Middlesex Hospital): Could Dr Barltrop tell us how his sample of children was selected for the demonstration of the distribution of blood lead levels? Was this a hospital population or a random sample of normal children from the community?

Dr D. Barltrop: The data were obtained from a hospital population of children who were in-patients with no known abnormal exposure to lead. There have been few attempts to determine the distribution of blood lead concentrations in a random sample of children and the findings would, of course, merely indicate the prevailing environmental conditions for the population selected for study.

---

* Kehoe, R. A. *J. indust. Hyg. Tox.*, 1940, **22**, 381; *Ibid.*, 1943, **25**, 71; *J. roy Inst. publ. Hlth*, 1961, **24**, 101–120a, 129–143.

General Discussion

Chairman: Professor T. P. WHITEHEAD (Queen Elizabeth Hospital, Birmingham)
Reporter: W. M. CATCHPOLE

Professor T. P. Whitehead: I welcome you to the discussion of this symposium and I should, first of all, explain that my expertise in this subject is virtually nil. I was interested in one or two papers recently because I think I can claim to have quite a collection of lead poisoning cases. I have published articles tracing lead poisoning to lead nipple shields, home-made wine, and golliwog badges[1, 2]; this is really a very brief history of my study of lead poisoning. I can also, I think, claim to be something of a conservationist, and this again brought a smile to my lips when I read through some of the material. I was a conservationist in the 1950s when it was not very popular, well not as popular as it is now, and along with the Editorial Board of *The Lancet*, and one or two other friends, ran a campaign against the use of arsenic for spraying potato crops, that is for destroying the potato haulms. I do not think our activity in stopping it attracted much attention at all. You may be interested to know that the farmers in the U.K., in large numbers, were spraying enough arsenic on one acre to kill 20,000 people. They killed one person, they killed hundreds of cattle, and it produced very little press comment or support for our campaign. It is a rather interesting swing in public opinion that one has seen in the last few years.

Now I have the job of acting as a referee this afternoon and it is very difficult to structure a discussion that goes on for a length of time, but I think that if we just dodge about all over the place it is likely to lead to confusion. I will try to structure the discussion in this way: first, we will talk about sources of lead and then I think we ought to hear a little from the petroleum industry about the amount of lead in petrol and the issues for and against taking it out, because I believe there are many here who do not know about these issues.

Then there are several people who, I know, want to discuss the measurement of lead in the air, and possible other sources of lead besides petrol. I think that discussion should follow on measurements in the human body and the significance of these measurements. This obviously leads us into a much more general discussion but, finally, I hope to point it towards what members of this meeting feel about the future and the areas where further studies should be made. In other words, can we get out of this meeting some indication of those areas which really need to be studied?

I will ask Mr Catchpole to lead off by making some comments on Professor Lawther's paper.

Effects of Weather on Urban Lead Levels

W. M. Catchpole (Consultant): I was interested in the apparent seasonal variations in lead in air presented in Tables I and III of the paper by Professor Lawther and his colleagues, since it would not be expected that emissions in Fleet Street would vary greatly from one weekday to another. The obvious potential explanation is variation in weather conditions, particularly rainfall and wind speed. The London Weather Centre is not far distant from the sites used by the MRC Air Pollution Unit. Table A presents some figures extracted from their records and compares them with the lead concentrations given in Table I of Professor Lawther's paper. Complete correlation cannot be expected because the weather records for the whole month have been used, as opposed to working weekdays in the MRC unit report.

The first column of weather information gives the rainfall for the month between 09.00 and 21.00 hours; the second gives the number of days on which no rainfall (not even trace) was recorded on the whole 24 hr of the day. Columns 3, 4, and 5 indicate respectively the average wind speed over the whole month in knots, the number of hours when wind velocity was over 10 knots, and the number of days on which velocities of over 10 knots were recorded.

TABLE A

	Pb μg/m³ 8–7 weekdays		Rain		Average	Wind	Days
Date 1971	Fleet Street	Med. College	9–21 hr mm	Dry days	speed knots	hr > 10 knots	with > 10 knots
April	4·9	0·9	14·2	18	10	338	26
May	6·2	0·9	35·6	15	8·1	160	23
June	5·4	1·0	16·9	16	10·2	330	24
July	4·9	1·0	17·1	17	8·4	221	21
August	5·6	0·9	20·4	11	9·8	319	27
Sept.	8·7	1·7	6·4	24	6·9	80	13
Oct.	8·6	2·3	23·8	23	9·9	314	19

It will be seen that there is a superficial relationship which suggests that high lead concentrations occur at periods of dryness and/or low wind speed, *e.g.* May gave higher lead contents than April or June and had the least hours of wind over 10 knots in these three months. September and October 1971 were remarkable for the number of rainless days and September had the least wind. This suggests that dry weather and low wind velocities cause an accumulation of inorganic lead in city streets, which may remain until it is either washed away or dispersed by high winds. To obtain a better understanding of the manner in which weather affects lead concentrations in city streets having little daily variation in traffic density, it would be necessary to compare results with weather day by day rather than monthly averages and also take into account street washing, demolition activities in the area, etc.

Professor P. J. Lawther: I am most grateful for that intervention. We would like to have your table and would like to give you the figures that I showed because we now have three more months. It is interesting that you should have mentioned demolition operations. Recently, when we had a very "upper crust" meeting on lead to get all latest results, I, as usual, had my Leica looking for lantern slides. During this meeting I got two photographs out of the window, near the Elephant and Castle, of exactly what you described—enormous bonfires, clearing old derelict houses, and I think this was making a considerable contribution. We have the daily results, some that we have obtained by high volume samples, and others by pooling weekly figures, and I am sure that Dr Commins would be delighted to supply these so that we could work out these correlations further.

Dr D. H. Peirson (AERE, Harwell): I would like to comment on Mr Catchpole's contribution about the possible seasonal effect. I think this is well recognized, not only in the case of lead concentrations but of other trace elements as well. I think the explanation is that in the summer, when the ground is warmer, the vertical diffusion upwards in the atmosphere is greater than in winter. In the winter there is a much greater frequency of inversions and less disturbance of surface level, so that one would expect the surface air to have a greater burden of material in winter than in summer.

Methods for Analysis of Lead in Air

Dr R. Stephens (University of Birmingham): With respect to the question of aggregation of particles, has Professor Lawther considered the possibility that aggregation occurred as a consequence of the sampling procedure rather than an aggregation in the air being sampled?

Professor P. J. Lawther: We have considered this. My colleague Dr Ellison could deal with the aerodynamic aspects. We do think that the aggregates are the kind which we see very frequently resulting from almost any combustion process. If we look at the smoke from burning benzene, we find long chains of carbon particles. We thought that the mechanism by which these chains might have arisen was that they could form at the flame front as it moved towards a cylinder wall and was quenching and that the lead condensed on some of these particles and was ensnared in others, accounting for the electron density.

Dr J. McK. Ellison (St Bartholomew's Hospital Medical School, MRC Air Pollution Unit): We have, in fact, used two types of sampling mechanism for electron microscopy, although only one type was shown on the slides this morning. The ones you saw this morning were entirely sampled on grids mounted in thermal precipitate, *i.e.* the ordinary standard thermal precipitator, and we do agree there is some possibility that some of the particles get formed in the sampling procedure as the sample is deflected towards the cooled surface, particularly as the thermal gradient does get slightly modified near where something is already deposited. Nevertheless, some of the pictures show a relatively small proportion of large aggregates. The other type of sample, we took only one or two, was on a membrane filter. That draws air straight up on to the surface of the filter. With this method it is very difficult to get a good electron micrograph because of the texture of the filter and the need to get rid of the filter itself before one can obtain an electron micrograph. I do not think there was any notable difference in the general appearance of the resulting pictures.

Dr R. M. Hicks (Middlesex Hospital Medical School): One important point that came out of Professor Lawther's presentation is that maybe all the lead levels measured are too low. It was quite clear that the nitric acid he used for extracting the lead from his electron micrograph samples did not dissolve all the lead and I believe I am right in saying that lead bromide, for example, is not soluble in nitric acid. The starting point for all estimations of lead in air is to dissolve your lead in nitric acid and then to estimate what has gone into the nitric acid. Well, how much of the lead is not being analysed?

Professor P. J. Lawther: In order to preserve our electron microscope grids, we used 20 per cent nitric acid, cold, whereas the environmental samples were stewed in boiling nitric acid.

Threshold Limit Values (TLV)

A. Cluer (Burmah Oil Co.): Can Dr McCallum enlarge on the basis whereby TLV values for lead have been selected? In the U.K. and U.S.A. the figure is 200 μg/cu. m, but in the U.S.S.R. the figure is stated to be 10 μg/cu. m. The values obtained in Fleet Street by the MRC Unit range up to 9 μg/cu. m. Set against the Russian standard, these would give cause for concern, but by the British standard we could be complacent.

Dr R. I. McCallum: The TLV adopted in Britain is the American one published by the American Conference of Governmental Industrial Hygienists. I have no knowledge of the data on which either the American or Russian figures were based. Possibly the U.K. figure should be regarded as a ceiling rather than a limit which could be exceeded at times.

S. G. Luxon (Department of Employment): TLV figures are limits for exposure of workers in industry for a 40-hr working week and must not be confused with occasional exposure.

G. V. Coles (British Rail): The American Conference of Governmental Industrial Hygienists has already proposed[3] that the TLV for lead in air should be reduced to 0·15 mg/cu. m, partly on the basis of work in the U.K. by Dr Williams and his colleagues.[4] The current level of 0·20 mg/cu. m was shown to correspond to a lead in blood level of 70 μg/100 g in an exposed population, whereas a TLV level of 0·15 mg/cu. m resulted in a blood level of 60 μg/100 g. Statistical analysis of their results indicated that a TLV level of 0·15 mg/cu. m was safe, whereas 0·20 mg/cu. m was not.

Professor D. Bryce-Smith (University of Reading): Russian TLV levels are based on disturbed behaviour in animals. 0·7 μg/cu. m is the lead level in cities.

N. C. Wildblood (Blythe Colours Ltd): TLV of 200 μg/cu. m referred to exposure at work for about a quarter of the 168-hr week. If one took a quarter of the TLV as the limit for continuous exposure, one would have a figure of 50 μg/cu. m for the existing TLV or 38 μg/cu. m corresponding to the proposed new TLV of 150 μg/cu. m. If children should not be exposed to more than a quarter of the lead burden that adults could stand, one would arrive at an equivalent of about 10 μg/cu. m—the Russian figure.

Professor P. J. Lawther: There is confusion between TLV and blood lead levels, and a need to establish proper relationships.

Professor T. P. Whitehead: Should one not examine blood lead levels of all people in a lead dust environment?

S. G. Luxon: Earlier speakers are confusing TLV for industrial exposure over a 40-hr working week with non-comparable continuous exposure of a non-industrial population. Periodical medical examinations of industrial workers are not dependent on airborne lead concentrations but are tied to certain specific processes involving lead.[5] They may be regarded as part of a dual-control procedure and complementary to determinations of lead in air.

Dr R. Stephens: Comparisons between airborne lead levels in a "lead smelting" or industrial area and from engine exhausts in a city street could be most misleading because of the difference in particle size distribution. The percentage of "active" or retained lead could be quite different and could make the car exhaust far more dangerous than the other. Expressed another way, TLV could have a particle size function. Secondly, one should not use stationary air concentrations to emphasize the small percentage of ingested lead unless one can determine the ultimate fate of airborne lead derived from leaded petrol.

Sources of Lead

Dr R. M. Hicks: The general view is that leaded petrol is the major contributor to lead in air. Data published by Fussell[6] and the National Academy of Sciences, Washington[7] indicate respectively that 90 and 98 per cent of the lead in air comes from this source.

Professor Lawther has found that a substantial proportion of the lead in the atmosphere on the Medical School roof comes from some source other than petrol. He has not extrapolated this finding to other localities and it is important that nobody should do so. For example, GLC figures for lead in air in St. Clement's Lane, which is near Fleet Street but has little traffic, are lower than on the Medical School roof. Professor Lawther's own figures show that the lead in Fleet Street has gone up 60–70 per cent in nine years, and the GLC traffic flow has increased by about 30 per cent in Fleet Street in the same time. When you compare this with extra congestion and the "revving up" that Professor Lawther mentioned, it seems quite plausible that the increase in lead came from exhaust fumes.

I live within 100 yards of the Cromwell Road where, according to the GLC, the traffic flow is the highest in the country. It is about three times higher than in Fleet Street so the lead levels may be three times higher.

Professor P. J. Lawther: We would all like to know not only the nature of the lead but the amount of lead in the part of the world you live. We have not taken samples but would be delighted to lend you a pump because we are not only measuring lead in different places but indoors and outdoors because, as you probably know, there is evidence that there is considerably less lead indoors than outdoors. With respect to our sampling site, as I was at pains to point out, we have been doing the measurement for research, not monitoring, and are trying to use the other pollutants as tracers to indicate whether there are sources of lead other than cars. We still have two-thirds of the lead on the Medical College roof unaccounted for. I do not know whether there are major sources near our particular site but I do not think anyone would be surprised. At other sites the motor vehicle is the major source and there might not be any other sources at all. We are talking, I think, on the assumption I see running through so much of the literature, that we are dealing with homogeneous populations. We are not, we are dealing with point sources all over the place; some are moving sources and some are very complex moving sources. I did hope to have made the point about motor vehicles in Fleet Street that it is not so much the number per hour or the density, it is a highly complex parameter but I would hate to give the impression that we were extrapolating from our work.

Professor T. P. Whitehead: I have the impression from papers that I have read, perhaps I do not read the right papers, that those who have done observations think that the majority of lead is from sources other than petrol. This is just an impression—who made the statement "the majority of", I do not know. Dr Hicks used the word "authorities" but implied that the majority of workers in the field would consider petrol to be the main source of lead in the air. There is a little confusion in my mind as to the true position. I know it will vary from place to place, but I would like comment.

Dr R. M. Hicks: Living right next to this road I can see what the traffic is. There is a jam extending from Hogarth roundabout in three directions for about one and a half hours morning and evening and when the M3 to Portsmouth is completed we are going to have it for 12 hours a day.

My comment about petrol as a source of lead comes from the U.S. Academy of Sciences Report, pp. 31 and 32.[7] They say that 98 per cent of airborne lead can be traced to its source, the combustion of leaded gasoline, and Dr Fussell has given a figure of 90 per cent.

Professor P. J. Lawther: I think one has to examine the data on which these claims are made. When you actually get down to the literature the amount of "guesstimate" in this field is unbelievable. I think that very few attempts have been made to estimate the actual source other than by kicking around the figures from the calculated emissions. Quite obviously, there are places in which the majority of lead is not from motor vehicles, for example, in the vicinity of a lead smelter or of burning old paint. In the middle of Fleet Street, I have little doubt that the majority of lead is from motor vehicles. We have already said that so far as our technique is concerned, the lead on the roof of the Medical College is only as to one-third due to vehicle emissions. I think one has to be very careful not to make calculations on emissions rather than airborne concentrations and the difficulty is that we have just not enough measurements.

Dr A. Berlin (Commission of the European Communities): From Professor Lawther's experimental work it is possible to estimate that nearly one-half of the lead in the atmosphere on the Medical College roof has an origin other than motor traffic, and must be connected with the "general particulate matter" in the atmosphere. Using the data presented in Table I of Professor Lawther's paper, one can calculate that the lead content of this "general particulate matter" would be about 0·4 per cent. The findings of the 1929 Ministry of Health Departmental Committee on Ethyl Petrol,[8] quoted by Dr Egan, seem to agree with this figure. Using this value, one can estimate that in Fleet Street the 347 μg of smoke would contribute 1·4 μg of the lead burden in the total of 6·3 μg/cu. m. Thus in Fleet Street at the present time, about 80 per cent of the lead seems to be due to motor traffic.

Using this same reasoning, the data presented in Table IV of Professor Lawther's paper seems to indicate that between 1962 and 1971 the share of lead pollution of the atmosphere in Fleet Street due to motor traffic has increased from 50 to 75–80 per cent. As to the source of the remainder, we were told this morning that the lead content of British coal averages 17 ppm (many coals are known to have a lead content between 10 and 40 ppm). Taking this figure and assuming that fly ash emission is 5 per cent of the quantity of approximately 150 million of coal burned annually in Britain, one can see that about 150 tons of lead is emitted each year from chimney stacks. This may seem small compared with the 6000–7000 tons of lead from motor fuel as indicated in Mr Stubbs' paper, but because of the altitude at which it is emitted it should not be neglected.

From the various presentations I came to the following conclusions.

—In cities, at street level, most of the lead pollution comes from motor traffic.

—In industrial areas, and especially above street level, lead from other sources, particularly fuel burning, becomes significant.

—Pollution of the soil by lead from motor car exhausts is significant up to a distance of 100 m from motorways (Dr Egan). This influence probably extends further.

—The lead transported over long distances, e.g. that found in the ice of Greenland and which started increasing significantly from the beginning of this century, is contributed by many sources, of which fuel and coal burning are more significant.

Dr H. Egan: The substantial literature on measurement of lead in or on pasture adjacent to highways suggests that lead levels (in the sense of ability to measure traces of lead) may be significantly different from remote areas at distances of 50, 100, or occasionally

150 m from the highway. The distances and lead level will depend on many local factors, including traffic density, speed of flow, prevailing wind direction, and wind speed.

Dr L. H. P. Jones: The contribution of motor traffic to lead content of plant growth has been overstated. In Oxfordshire, the lead content of grass reaches a constant level at 30 ft from a road used by 12,500 vehicles/day. There is no difference on a doubling of the traffic density.

Professor P. J. Lawther: I am grateful to Dr Berlin for these estimates. My mental arithmetic is not as quick.

Dr B. T. Commins: I too thank Dr Berlin for his comments and calculations. I think it is extremely difficult to assess what is the contribution of lead from sources other than traffic. We have a tremendous amount of information, but we need a lot more, and all the weather data, before we can answer some of the questions. The results so far indicate that on our roof there are substantial amounts of lead from sources other than traffic. One thinks of other sources and maybe incineration plants contribute some. We will continue to take measurements on our roof and hope to be able to provide further information.

The Need for Lead Alkyls in Motor Fuels

Professor T. P. Whitehead: I was puzzled by one or two points while reading up the subject before taking the chair, some of which might be explained by representatives of the petroleum industry. For example, I am unclear what happens to the lead which does not come out in the exhaust. Does my car get heavier every year? If it goes into the sump oil it must go somewhere afterwards—where does it go? I am a little puzzled because 25 per cent of 9000 tons is a large amount of lead, and I would like to ask where it goes.

J. D. Savage (British Petroleum Co. Ltd): Some of the lead stays in the combustion chamber, some passes into the lubricating oil in the sump, and some is accumulated in the exhaust system. Lead is used to give anti-knock performance and to reduce wear. It is not essential but most present-day engines are designed to use leaded petrol. Remove it immediately and the result will be power loss, engine damage, and increased consumption. In the U.S.A. a lead has been given to make use of unleaded or "low lead" fuels feasible so that catalytic converters can be used to reduce other undesirable emissions. In Europe steps have already been taken by some countries to reduce lead. If lead removal can be justified both ecologically and economically it will be necessary to set a dateline sufficiently far in the future to enable compatible fuels and engines to be developed and for existing ones to be phased out. We can benefit from the action and the mistakes which have been made elsewhere and see that the oil and automotive industries work together towards satisfactory emission control.

J. H. Boddy (Mobil Oil Co. Ltd): We have heard from Mr Stubbs that petrol represents only 9000 tons of the 272,000 tonnes used annually in the U.K. Is this 9000 tonnes more of a hazard than all the rest? More information must be obtained to resolve this uncertainty but, as stated as recently as yesterday in the House of Commons by the Minister for the Environment, it seems certain that there is no positive evidence to indicate that lead from gasoline represents a health hazard in the U.K. at the present time. Nevertheless, the Minister also stated that he was advised that it would be prudent to have increasing lead consumption and, if possible, to reverse the trend.

  We now have a position where cars developed for high efficiency and fuel economy could not operate without lead, which is the most flexible way of providing anti-knock requirements in gasoline. Given time to re-equip refineries at a cost of £200–250 million, we could eliminate lead but the gasolines produced would not be as high in anti-knock rating as at present, and we would use more of it less efficiently. This would entail higher imports of crude oil and this is probably more significant to the national economy than the fact that consumers would have to pay up to 2½p more per gallon because of the higher manufacturing cost.

  Just as this conference is reviewing lead in the environment from all sources, so do I believe that all emissions from vehicles should be considered in conjunction with lead, so that we can make sure that reducing the level of one pollutant will not increase others equally noxious. It is the policy of the oil industry to co-operate fully with the automotive industry and with governments. Through national and international working groups such as the British Technical Council of the Motor and Petroleum Industries, the Co-ordinating European Council for the Motor and Petroleum Industry, and CONCAWE, relevant information on emissions, engine performance, and design is being collected and made available so that a balanced assessment of the best way to minimize pollution may be reached.

Dr. R. M. Hicks: I query the ethics of an industry which, according to Mr Boddy, will say that it is not a good economic policy to put up the price of petrol by 1 or 2 pence per gallon in order to omit lead additives when there is any possibility that lead in air from exhausts adds to the lead burden to which young children and the unborn fetus is exposed.

Dr D. Taylor (Imperial Chemical Industries Ltd): In reply to Dr Hicks, the ethics of the petroleum industry in manufacturing a potentially harmful substance can perhaps be compared with the ethics of the motor industry and car users as a whole. Sir Eric Ashby, in his opening remarks this morning, pointed out that we seem quite ready to accept the deaths of several thousand people on the roads each year without qualm; this is a valid comparison.

As a question to Mr Boddy, what percentage of capital presently invested in oil refineries is represented by the extra £200–250 million to produce lead-free petrol?

J. H. Boddy: I cannot give a precise answer as I do not have the figures here but I believe they could be made available.

A. Cluer: The cost of a modern refinery to make fuel products is roughly £10 per annual ton of oil throughput. The refining capacity of the U.K. is at present about 100 million tons/year, representing an investment of the order of £1000 million. Thus the expenditure mentioned by Mr Boddy represents 20–25 per cent of current investment.

I should like to add, while considering such vast expenditure, that the oil industry most certainly recognizes its responsibility to the community. None of us wishes to be choked or poisoned out of existence or to see this happen to our children. However, on the evidence so far presented against lead in motor fuel, it seems that we as a nation could find better ways of spending £200 million to clean up the environment than by omitting lead, for example, in cleaning up water supplies.

C. L. Goodacre (Consultant): I cannot believe that lead in the environment from motor car exhaust is the hazard to health that Professor Bryce-Smith and others tell us it is. It seems that the findings to date of the Medical Research Council presents the true situation. However, if, repeat if, a reduction of lead emitted to the atmosphere from motor cars is or becomes desirable, then the following action could be taken without developing any new technology. It seems, in the long run, it could be both a financial and material "life-saver" for all concerned, in particular the customer, to give consideration to the following points:

1. Tetraethyl lead (TEL), as an anti-knock for motor gasoline, should be progressively and quite quickly phased out in favour of the more effective tetramethyl lead (TML) anti-knock, for the following reasons:

(i) No uninhibited petroleum technologist, dealing with motor gasoline manufacture today, would now freely choose TEL as an anti-knock, boiling at 200° C, at the end or beyond the end of the gasoline's boiling range, if he could use TML, boiling at 110° C, in the mid-point area of the gasoline's boiling range, if there was not an initial "purchasing" bias in favour of TEL to obtain CFR laboratory knock test engine octane numbers (Table B). Motor gasoline is made to propel motor cars along the road, and not to run laboratory knock test engines. This point is sometimes set aside for commercial reasons.

(ii) A mass of data exists to show that TML gives equivalent road octane numbers and pre-ignition rating for "less weight of lead metal" than TEL. The "road advantage" for TML over TEL can be 25 per cent less weight of lead metal for equivalent road result, in particular in European-type motor gasolines in European and Japanese cars.[10]

(iii) When TML was proposed by Calingaert et al. of Ethyl Corporation, some 40 years ago, as a superior anti-knock to TEL for multi-cylinder engines, automotive and aviation,

TABLE B
Lead Alkyls Physical Characteristics

	TML	TEL
Boiling point at 760 mm Hg	110° C (230° F)	200° C (392° F) approx.
Molecular wt.	267·4	323·5
Metal content	77·5% wt.	64·06% wt.
Density at 20° C (68° F)	1·995	1·650
Melting point	−30·3° C (−22·5° F)	−130·2° C (−202·4° F)
Refractive index	1·512	1·520
Vapour pressure (mm Hg)		
at 15° C (59° F)	17·50	0·167
at 50° C (122° F)	100·00	2·10

the use of TML was forbidden on the, now known to be fallacious, premise that TML, being 50 times more volatile than TEL at day ambient temperatures, would be "deadly" by inhalation. The basis of this premise was the extrapolation of TEL experience to TML. This premise was proved to be completely wrong. Investigations between 1956–59 suggested that the human health hazards from TML were substantially less than with TEL, by a factor of about 7; this convinced the U.S. Surgeon General that TML was a marketable product, despite its much increased volatility.

(iv) It appears from the literature that TEL intoxication is primarily related to the breakdown of this lead alkyl to the tri-ethyl lead radicle in humans, with the consequent rapid agglomeration of lead in the body.[11] It appears from the literature that TML does

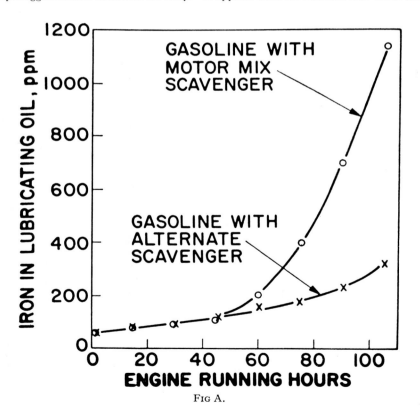

Fig A.

not break down in the human body, *i.e.* it goes in by inhalation or absorption as TML and comes out as TML; thus it does not build up lead retention. Strange as it may seem, laboratory tests of TML–TEL confirm this point, as TEL starts decomposition at 100° C, TML at 265° C, TML being a much more stable material than TEL.

It would be very interesting now to review the yet unpublished work on TML related to TEL toxicity in the archives of the companies wishing to market TML in the U.S.A. and elsewhere in the 1956–60 period and onwards. The banning of TML 40 years ago

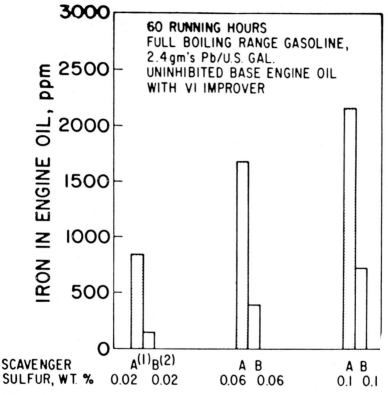

60 RUNNING HOURS
FULL BOILING RANGE GASOLINE,
2.4 gm's Pb/U.S. GAL.
UNINHIBITED BASE ENGINE OIL
WITH VI IMPROVER

SCAVENGER $A^{(1)} B^{(2)}$ A B   A B
SULFUR, WT. %   0.02   0.02   0.06   0.06   0.1   0.1

(1) MOTOR MIX–1.0 THEORY ETHYLENE DICHLORIDE/0.5 THEORY ETHYLENE DIBROMIDE.

(2) ALTERNATE SCAVENGER CONTAINING LESS HALIDE.

FIG B.

has probably cost the petroleum industry countless millions in installing, unnecessarily or prematurely, gasoline octane number upgrading equipment they did not need.

(v) The better distribution between the cylinders of an engine with carburation and manifolding, of the lead alkyl anti-knock, under idling, low power, and accelerating dense traffic conditions, obtaining with TML, as opposed to TEL, are well known and should result in more balanced combustion conditions for the lead alkyl, thus emitting less organic lead from the car's exhaust.

(vi) According to published data, sales of TML anti-knock compounds in 1970 from European sources amounted to some 37 per cent of the lead alkyl business, the balance being TEL. To phase out TEL in Europe could cost the oil industry some £6,500,000

6

in initial re-equipment in lead alkyl plants. Overall, this action could show, in short term, an ultimate saving to the producer, the customer, and the nation.

(vii) A survey of recent patent literature shows that there already exist "lead traps" which can be installed in the car's exhaust system, at low cost without serious operating penalties, as particulate traps on the well-proved cyclone principle. Claims are made that up to 50 per cent of the weight of lead metal ingoing to the engine can be retained in the trap device for 50,000 miles of operation, minor simple servicing being required at infrequent intervals, such as oil filter change periods.[12]

2. The organic chlorine compounds currently used in motor fuels as "scavengers" to reduce combustion chamber deposits should be phased out in favour of the di-bromo ethane used originally.

Scavengers of lead complexes developed in the Otto cycle engine's combustion process, when operating on leaded gasoline, rely on halogens based on bromine and chlorine. The initial choice, around 1924–25, was di-bromo-ethane boiling at $131\,^{\circ}$C, for TEL boiling at $200\,^{\circ}$C, the bromine being at one theory on a mole-mole basis. This TEL mix is known as aviation mix and has not changed in over 45 years.

Bromine production shortage in 1929–30 forced the replacement in major part of the bromine complex by di-chloro-ethane boiling at $87\,^{\circ}$C, as a "makeshift" scavenger for TEL as 44 (and later 62) motor mix, one theory of DCE being necessary with half a theory DBE, i.e. a one and a half theory motor 62 mix from 1943 to the present time.

Unfortunately, this cheaper and much inferior chlorine-based DCE scavenger has been found to wear out the engine in road service, at least at twice the rate of the bromine-based original (Figs A and B).

This accelerated engine decay has been well proven in modern cars by Hudnall et al. of Mobil Research and Development Corpn.[13]

Other workers in Europe have confirmed Hudnall's findings re the destructive effect of chlorine-based scavengers to the engine.[14] It would now seem desirable to remove chlorine from gasoline as quickly as possible. The faster the car engine wears out, the more the atmosphere is polluted.

Professor T. P. Whitehead: The points that you made were, first, that TML is not as toxic as many people have said; secondly, it would reduce the amount of lead.

W. C. Greaves (British Petroleum Co. Ltd): I would like to respond to Dr Hicks because I am not here today as a medical man or an engineer, I am here basically because I am worried about the ethics of the problem in the same way as is Dr Hicks. As she is basically asking me a question whether I am worried about the ethics of this sort of problem, my answer to Dr Hicks is that I have worried about it, I have read a series of reports about it over a long period of time. There was a report issued by the Government on ethyl when it went into this subject in about 1930.[8] There have been two major reports in the U.S.A., one by the Surgeon General[14] and one issued last year by the National Research Council[7] and we are having a discussion about the subject today. All of these reports, as I see them, have arrived at broadly the same conclusion as that at which we arrived this morning. Somebody questioned one of the speakers about the significance of the effects of the air-borne pollution on children, and the answer was that the effect was insignificant. He put up some figures which showed that, taking the most conservative estimates in the wrong direction, out of 156 μg of lead that went into a child's body over a period of time, 3 μg were absorbed from airborne pollution. As I understand from what Professor Lawther said this morning, the probable order of the contribution of motor fuels to that figure is 1 μg. Our industry problem in these circumstances is that we have listened to this sort of discussion and that lead from the atmosphere is a minor constituent in the total health problem. In this audience today the only dissentients appear to be Professor Bryce-Smith and Dr Hicks, and as far as I can see they have produced no evidence to contradict the opinions stated in authoritative reports and neither did Professor Bryce-Smith when I went to listen to him previously in his TV presentation. We are prepared to listen to whatever new evidence these people have and want to review. You heard Mr Boddy and Mr Savage describe the technical and economic problems. I will start with the £200 million that Mr Savage mentioned, or the figure from my company, which we have calculated to be £30 million for the U.K. The question for us, at this point of time, is how do we justify spending £30 million to take lead out of gasoline on the basis of the kind of evidence that has been produced this morning or the evidence produced by any previous report? We have argued about it. Any management of a responsible oil company must have argued about this in the light of the discussions over the last year.

We have discussed the cost of doing this with the Department of Trade and Industry. As far as our company is concerned, it looks like costing us one new penny per gallon. What I am saying to Dr Hicks is that if she can persuade the British Government and the British motorist that he should either travel at 30 miles an hour or that he must pay more for his motor gasoline we are prepared to accept that and remove lead. It is not an issue that we, or any single oil company, the oil industry, or the motor industry separately can take a decision about.

Dr P. S. I. Barry (Associated Octel Co. Ltd): I just want to make one point in relation to Mr Goodacre's remarks. He indicated that TML was virtually not toxic and it does not undergo a change in the body. This is not, I regret, correct. TML does undergo a change in the body, as does TEL, but it does it at a much slower rate. Cremer found that on a weight for weight basis TML is of the order of seven times less toxic than TEL.[15-17] Dr Kehoe's group in the U.S.A. in the early 1960s showed it to be of the order of four times less toxic.[18] It is not correct that tetramethyl does not undergo breakdown to trimethyl radical.

Effects of Lead on Health

Professor T. P. Whitehead: I am now going to move the discussion towards what one might call the medical aspects.

Professor D. Bryce-Smith: I think that it is important that those of us who have expressed concern about the present level of lead exposure to the general population are not presented as individual cranks who are raising a scare without any evidence and without support from the most authoritative quarters. Concerning the actual seriousness of the problem itself, this has been widely appreciated in America and indeed most of the paediatric studies which are of relevance come from America and they have quite clearly indicated that the problem reaches its most acute form in children. In 1969 the American Academy of Paediatrics issued an official statement on lead poisoning in childhood. I wish to quote briefly from this.[19] "Prospective surveys in various cities indicate that 10 to 25 per cent of children who live in deteriorated pre-World War II urban housing have absorbed potentially dangerous quantities of lead and that 2–5 per cent have clinical symptoms compatible with those of acute intoxication." Those are significant figures, I submit, by any standards. Furthermore, the criterion of undue exposure of children which Dr Barltrop gave us, namely, 100 μg/100 ml of blood, is not supported by this extremely authoritative report, which is based on screening studies of tens of thousands of city children in the U.S.A. They say that if you find a child, even an asymptomatic child, having a blood lead level of more than 80 μg/100 ml then that child should be rushed off to a clinic immediately and treated for lead poisoning, as it is in great risk of suffering lead encephalopathy. Other studies have shown that a high proportion, and it may be as high as 90 per cent, of children who, before the age of five, suffer an attack of clinical lead poisoning, not necessarily encephalopathy, suffer permanent brain damage of various sorts resulting in specific educational disabilities and behaviour and emotional disturbances, which are indeed of sociological as well as medical significance. So that is the measure of the problem as it is seen authoritatively in the U.S.A. at the moment. We have not, to my knowledge, initiated the sort of mass screening of city children in the U.K. that is being carried out in the U.S.A. and I would like to take this opportunity of calling publicly for the introduction in the U.K. of mass screening of city children for elevated levels of lead much along the lines now being carried out in the U.S.A.

Now I wish to turn to the health aspects of this business of lead in petrol. This meeting has been organized by the Institute of Petroleum and I imagine that there are a lot of people here who feel that this is one of the most critical factors to get clear in one's mind. There are two aspects of this problem, first, inorganic lead, and secondly, organic lead. Most of the organic lead in petrol ends up as inorganic lead. The proportion that does not seems to be generally unknown. Some very high figures were given by Professor Lawther this morning. I hope that they are wrong but generally the figures for the organic lead have indicated that the levels in city air are between 2 and 10 per cent of the inorganic lead level. Certainly, people have been poisoned by undue exposure to leaded petrol, *e.g.* by use as a de-greasing or dry-cleaning fluid, and the initial symptoms of poisoning are extremely difficult to recognize because they mimic those of what one might call a conventional psychotic illness. Nevertheless, this dangerous material is sold in the U.K. without any health warnings to garage mechanics, who still use it quite widely for de-greasing purposes in closed environments. The second thing that I am calling for, therefore, is that if the industry wishes to go on selling leaded petrol it should issue publicly to garage workers and the general public, if necessary through the AA, a warning against avoiding undue exposure to the vapours of petrol or using it for any purposes other than putting it in a car to drive along. Now I want to turn to inorganic lead. There has been concern expressed of the great disparity between the levels allowed in Russia and the levels allowed in Britain. The levels allowed for industrial exposure in Russia are one twentieth of those allowed in Britain, namely, 10 μg/cu. m, and these have been arrived at on the basis of experimental studies of disturbed behaviour in animals. No such studies have been carried out in the West. The levels adopted in Russia for the whole population exposure are in fact much lower than that, 0·7 μg/cu. m is the maximum level for exposure in cities, and virtually everybody in London these days, living anywhere near traffic, will be exposed to more than that. Finally, I wish to quote from an American report, the United States National Academy of Sciences report, "Airborne Lead in Perspective, 1971".[7] First, what Dr Hicks said about 98 per cent of the airborne lead that can be traced to its source comes from combustion of gasoline. Secondly, the amount of inhaled lead is about one half, and in special circumstances twice, the amount that comes from diet, and that is on p. 318 of the report. Next, there is no evidence of any health threat from the ambient levels of lead in air to the general population but there is evidence that special groups such as traffic policemen are exposed to airborne lead to an undesirable degree. Thirdly, they point out, as far as children are concerned, that many children with documented prior attacks of symptomatic lead poisoning develop hostile, aggressive, and destructive behaviour patterns which in turn may precipitate their exclusion from school and a demand for institutionalization. Fourthly, and finally, the level of airborne lead "poses a significant threat" to inner city infants and young children. In view of the evidence that nearly all of this comes from the practice of adding lead as an anti-knock agent to petrol, I feel that this report points the way clearly to the responsibility of the oil industry in this matter.

A. J. Lambert (Burmah Oil Co.): One of the things that strikes me from the opening talk this morning is to put things over as simply as possible to the public and certainly there is one item that I do not understand at all clearly, and that is that we have talked about lead and the increasing quantity of lead being used in gasoline, and so on, but I have not seen anywhere any way in which the ordinary man can diagnose when lead poisoning is about to occur. In other words, we have not got any simple diagnosis of lead in man, symptoms of toxicity. A common cold is readily understood, but I think that if one has something which was common cold symptoms it might in fact be lead symptoms. How would we know? Are the people of the country being poisoned by lead? Well there is no evidence of that so far, because one hears of no cases reported of lead poisoning to the ordinary human being, only perhaps to workers in industry, a few cases which one has to search for. For the benefit of myself, who does not understand this, and partly for the benefit of the general public, it would be worth perhaps having some general comments from Dr McCallum and Dr Barltrop and how they would put this over to the press and the public to say how they can diagnose, see the symptoms, appreciate what the toxicity of lead really is.

Professor T. P. Whitehead: Dealing with the early symptoms of lead poisoning, are there any that can be detected?

Dr R. I. McCallum: I tried to make the point earlier that there is great difficulty in deciding whether a person has, or has not, lead poisoning at a certain stage. It is, in fact, a clinical opinion, which has to be based on the history of somebody who is complaining, and the sort of measurements which I showed on the screen, the blood lead, the copraporphyrins, the ALA level in the urine, and so on. At one extreme we have lead poisoning of a traditional kind; in industry, the encephalopathy which is extremely rare and I have never seen it. It is so rare that only certain people in the petroleum industry have seen it. I believe it is associated with cleaning out tanks that have held petrol and I must make an obvious point, because I understood Professor Bryce-Smith to suggest that this was a risk to you and I as we have our cars filled with petrol at the filling station. I think that this is a gross exaggeration, if that is what he implied. The classical symptoms of lead poisoning are those of colic, *i.e.* a pain in the stomach and paralysis. These symptoms make the diagnosis a little easier, although you can run into trouble through confusing a stomach pain with some surgical emergency. When you get to the point where the symptoms are much less striking than that, you run into grave difficulties. An early sign of lead poisoning is anaemia, but you have to examine the blood to detect that. It is at this stage that people may, or may not, complain and it is at this point that it is very useful in industry to have regular examinations of the blood. It has been pointed out by Dr Williams that measuring anaemia is a very crude estimate of lead poisoning; it is a little bit late then. We want to stop people getting to that point but at an earlier stage any symptoms are indistinguishable from the sort of things that you and I suffer from, from time to time, and are so diffuse they are difficult to quantitate. I hope that answers Mr Lambert's question about diagnosis, that it is very much a matter of clinical opinion, based on all the evidence that can be obtained.

Dr R. M. Hicks: I question Dr Barltrop's conclusion that lead in air makes an insignificant contribution to the lead load in children. I think he established very clearly that children are at a greater hazard than adults to lead, because of the way that they interact with the environment. They are likely to ingest more from a variety of activities which are classified as pica. He made a calculation on what a child would inhale in an atmosphere containing 2 μg/cu. m; 6 μg daily intake and a retention of about 3 μg. I do not question this and if you are basing this on average hazard to the average child taken over the whole country, I do not think that there is any argument. I am not talking about average concentrations or average populations, I am talking specifically about the effects of high lead levels in the atmosphere on the urban child. His level of 2 μg per cu. m, which was the starting point, is low under these circumstances. In Fleet Street it would be about 6 μg/cu. m and the daily intake would be 18 μg intake and 9 μg retained. I live within 100 yards, in fact, about 150 ft of the Cromwell Road, where the traffic flow is, according to the GLC, the highest in the country; it is about three times higher than that in Fleet Street. So maybe the lead levels in the vicinity of my house where I live, bring up children, where there are schools, a road going through a playing field, and there is a maternity hospital in the same road, are three times higher than in Fleet Street, which is already regarded as fairly high. If my figures are correct and Cromwell Road is three times higher again than Fleet Street, then the intake could be 54 μg and the retention 27 μg. I cannot believe that Dr Barltrop would maintain that this would be an insignificant intake when it is superimposed on the load to which urban children are already exposed from soil, food, water, and everything else.

P. Draper (National Society for Clean Air): With regard to Dr Hicks, I suggest that she could rest assured that she need not multiply the Fleet Street numbers by six due to the fact that the traffic flow is six times greater on the Cromwell Road, because the point is that the Cromwell Road traffic is moving fast. It disperses its exhaust very well, for one thing; another thing is that the combustion will be better, so that the lead is more likely to be trapped in the exhaust system than it is in Fleet Street, where it comes out with liquid or nearly liquid petrol under very rich conditions. Of course, we can refer also to Professor Lawther's earlier statement about his zero flow. So I do not think that because there is six times as much traffic this lady needs to get six times as worried.

Dr R. M. Hicks: Living right next to this road, I can see what the traffic flow is and there is a traffic jam extending along the Great West Road from the Hogarth roundabout in three directions every morning for about one and a half hours and every evening for about the same time. Now that the M4 has been opened to Wales, the length of time of that traffic jam is extending, and as soon as the M3 is open down to Portsmouth it is going to extend a great deal more, so that instead of having near-stationary traffic just morning and evening we are going to have it for 12 hours a day, in which case I am afraid there your argument about fast-moving traffic does not apply.

Dr D. Barltrop: I take it that Dr Hicks means absorption rather than intake in her calculation? She was referring to retention in saying 27 micrograms, I think. But first, in reply to Professor Bryce-Smith, though far be it from me to rank myself against the American Academy of Paediatrics, and I have no intention of doing so, on the other hand I think that the figures that he has given do not in fact match our clinical experience. I would certainly stand by the figure that I gave this morning. I think that really answers the question as far as this is concerned. I think the point made again about the significance of individual blood lead levels, e.g. 80 μg/100 ml or above, which the Academy made, is a valid one. I think that these cases should be treated seriously if only because we do not know, when we find an individual value, whether this represents a static state of affairs, a plateau as it were, balanced between absorption and excretion, or whether the child's blood lead level is increasing or falling. Certainly I would treat such a child seriously. I think we would obviously want to remove him from his environment when we established what the source of lead was and what its magnitude was. I doubt very much whether we would institute treatment, unless there was symptomatic poisoning. That is a point that I think would be best discussed again. Concerning the question of children with bizarre behaviour, aggressive tendencies, troubles at school, well of course they occur in children, many of whom have not been exposed to lead. Certainly my readings of the report left the impression that we really did not know whether sub-encephalopathic exposure, if that really is the right term, is significant or not; that is to say, we do not know if asymptomatic children with blood leads in the region of 40–80 μg/100 ml really are at risk or not. Obviously there are conflicting views on this but I think this is open to discussion to a considerable degree and I am sure some more is going to be generated. A point that I think I would make with both Dr Hicks and Professor Bryce-Smith is that the sort of thing that we are talking about in terms of micrograms of ingestion are absurdly small compared with the other sources of lead which I think they both recognize, and which I think, as a paediatrician, we should be diverting our money and our energy to do something about, namely, lead paint in old homes and lead in soil. Surely we should deal with greater hazards first and the lesser hazards second.

Professor T. P. Whitehead: From several speakers this morning it was obvious that there could occur in our cities, and perhaps outside the cities, what you might call "hot spots" of lead in the environment. You can imagine a combination of quite an aged house with burning rubbish near and the right traffic conditions. You find that blood leads in children give you a beautiful Gaussian distribution and then you get one or two children who are right outside this distribution and who could possibly be associated with "hot spots". Do you feel that a study of this kind is worthwhile in our cities? I think this is not in order to trace the effect of lead in petrol, although that would be part of it, but to see generally if the environment is producing "hot spots" with abnormal levels.

Dr D. Barltrop: I would certainly support screening programmes in children, if only to identify those children who may, or may not, be exposed to particular environmental hazards. Again, referring to my paper this morning, I see much more point in trying to identify the environmental hazards first rather than directly using the intermediary of the child. It seems absurd to me that we use children as indicators of environmental hazards. I think that we should go and identify the hazards directly. I certainly advocate and I would support massive screening programmes because they would be far more important than some of the things that we do.

It means a very large number of children indeed, if we look at this over the whole country. Even in the U.S.A., where this work is carried out, it is only in New York and Chicago, two major centres. Some others are coming along now, but really it is only in a restricted number of cities where this sort of programme is actually taking place.

Professor D. Bryce-Smith: I think there has been some misreading of the report; this is really a fundamental aspect of the whole American question. The question, I think, raised in this report is not, as Dr Barltrop had thought, whether children who do not suffer encephalopathy have permanent brain damage and other unpleasant sequilae;

rather, the report expressed uncertainty whether the children who have high blood lead, but no symptoms at all, suffer these very dangerous consequences; this is the point. There is some indication from the work of Perlstein and Attala,[20] in which they did follow up some children who had been found accidentally to have high blood leads but were exhibiting no abnormal symptoms at all. About 10 per cent of these children showed educational disabilities when they were followed up in later life. This question is not proved, and is possibly the most important question in the whole issue.

Professor P. J. Lawther: I just wanted to make one small point. I think that at the end of a symposium which has been as valuable and as stimulating as this, it would be regrettable if we saw arguments and opinions being polarized. We are all here as compassionate men and women and we are all after the truth. I think that it is sad if we persist in assuming levels, such as 50 per cent absorption, when after back-breaking work we have given you some evidence that this probably is not so. If we, in a learned society, are bashing our heads on the wall I think the cause is rather put back a little.

Dr J. A. Bonnell (CEGB): Some years ago, I was, in fact, concerned with lead poisoning and I would like to endorse what Dr Barltrop has said. Invariably all these children have suffered from lead due to ingestion. In point of fact, Professor Bryce-Smith earlier referred to the report to the Academy of Paediatricians of the U.S.A. These were children, as I understood it, from old homes; in other words, they were suffering from lead due to ingestion and not from inhalation.

Dr A. E. Martin (Department of Health and Social Security): At the beginning of this session the Chairman said that towards the end of it he would be introducing the subject of fields in which more research was needed. Earlier in the morning session, Professor Lawther referred to what he called an "upper crust" meeting in the Department of Health; the purpose of that meeting was to outline some of the areas in which research was needed. As a result, the monitoring of children has already been stepped up and we have seen the results coming through in the case of the Isle of Dogs. We are thinking also in terms of Dr Barltrop's work on paint and other "hot spots". On monitoring of lead, Professor Lawther has indicated the need for increases in these services.

There is one particular point that I would like to discuss and that is ALA dehydrotase, which was mentioned this morning, when it was said that there was difficulty is assessing its importance. The group discussions which took place in the Department of Health indicated that, so far, it has not been possible to correlate any adverse effects on health with this test. Dr Martin Holdgate has drawn my attention to a very interesting article on ALA-D which has recently been published.[21] Three groups of dogs, one as a control, one to which lead was administered for a period of a month, which reduced the enzyme activity to 50 per cent, the third group to an extent which virtually abolished the ALA dehydrotase activity. At the end of the month, blood examinations showed no difference between the three groups. They were then bled to one half of their blood volume and allowed to regenerate their blood. The results of that showed no difference in the regeneration. This again indicates that the body probably has adequate reserves to deal with this enzyme inhibition. This is only one subject on which we need much more information before we can be absolutely certain. Over the whole field of lead we need a lot more information. It is coming in but we still have no evidence of any community hazards, although there are hazards in certain hot spots, as has been mentioned today.

Dr J. R. Glover (Welsh National School of Medicine): I can describe a metal survey that we are doing in the Swansea Valley. The Swansea Valley, I should say, has no indigenous lead but lead smelting has been carried on there for 100 years and in this survey we are doing 11 elements in the blood of volunteers at six centres. The 11 elements will include lead, mercury, arsenic, copper, nickel, vanadium, selenium, cobalt, and zinc; there are two others. Dr Peirson, from Harwell, who is here, will be doing the air pollution at these six centres and we have two control centres. Dr Elward, from the Epidemiological Unit of the MRC, will be taking 50 women from the electoral roll in each of these centres (we cannot use men because they go away from each centre to work during the day), 50 children, and probably 50 toddlers, aged 0–5. Dr Martin pushed us into the toddlers. We are only doing blood samples for reasons of the difficulty of the whole project.

We are taking part of the Gower Peninsula as one control group, Kidwelly as another. The prevailing wind is westerly. We are taking our six centres at Cwmbach, Pontardawe, Llansamlet (where horses have died in the fields due to lead poisoning), Skewen, Tonna, and near Port Talbot, at the steel works there. Also co-operating in this survey is Dr Goodman, from the Department of Botany, who really started off by growing mosses in this area and showing[25] high quantities of lead and other elements which mosses concentrate. In fact, a moss desert exists in this area because of the lead smelting.

We are doing plants, soil, animals, and milk in co-operation with the Ministry of Agriculture. The important thing that we have really been discussing to date is our various interests. We insist, on the medical side, that we must not be within 100 yards of a road, otherwise we will just be doing what this conference is trying to do, measure whatever lead is coming from the road. We must, however, be within, say, 300 yards of a population of 500 people. We cannot, as Dr Peirson would like to do, go up the local mountain and take air pollution, because there we will not have a population to analyse. It is extremely difficult to get all our interests served. We hope to do this survey in the summer and I will be grateful for any suggestions or criticisms of this work. I was not expecting to put it to this audience but so many people have said that we ought to do more in this direction. I feel a terribly important thing in this particular survey is that we are not just looking at lead, we are looking at other elements and we are covering the air and the diet.

Dr W. C. Turner (London Borough of Tower Hamlets): I have just been involved in such a survey in the Isle of Dogs, as many people know, and there are some very interesting points coming out of it. One of them is that we have to think not in terms of what the cars put out as lead, because although this is undoubtedly in the environment, a lot of it, practically none of it gets into children. We have been studying a particular "hot spot" and things are rather different there. One of the interesting things is the correlation between pica and absorption in children who are grubby.

In the epidemiological study on the Isle of Dogs, within a natural perimeter at a radius of 500 metres from a long-established lead works, the blood levels of 157 children under 5 were compared with their mothers. Several relationships appeared.

The majority of raised blood leads were found in children living within a few yards of the main road which runs round the Island. This coincided with the highest deposited lead at the points of intersection of an 1/8th mile grid placed over the area in an early phase of the survey. Lead from petrol vehicles is not thought to be particularly significant, as the road is not heavily trafficked and most of the vehicles are buses or heavy diesel lorries, but it is thought to relate to deposits on the highway from vehicles delivering lead-containing wastes and the carry-out from the factory of lead-containing slurry on the wheels of vehicles and the feet of workmen. The slurry arises from the damping-down within the works environment to eliminate dust as far as possible.

A considerable number of these children were described by their parents as getting very grubby during play. Many were cases of pica. The child with the highest lead was a case of pica (soil in potted plants) living at 500 metres to the NW of the works. Repeated blood tests on this child showed that the blood lead figure was extremely labile.

The next four highest blood lead levels (75, 74, 65, 57) were found in households where the father was employed in the lead works and this held when they resided outside the survey area. The first three were admitted to the Hospital for Sick Children, Great Ormond Street, for full clinical investigation. The blood lead level fell rapidly and rose slightly when they returned to the home environment, even after special cleaning of the accommodation. There was no evidence of clinical involvement in any system or any question of lead poisoning.

Recent work in America, using micro-analytical and electron microscopy studies, supports the possibility of an adaptive mechanism in the human body. This is not surprising in view of man's continuous exposure to lead in the environment during evolution. At low levels of blood lead there appears to be a linear relationship between intake, blood lead, and excretion, predominantly as inorganic sulphate and phosphate. As the blood level rose, there was an additional excretion of organic lead which ultimately represented 40 per cent of the total excretion. At the same time there appeared in the nuclei of hepatic and also of renal cells in the proximal convoluted tubules, inclusion bodies shown to be non-diffusable lead protein complexes which increased in size and number until there appeared evidence of progressive renal damage with abnormal clearances. In experiments on rats, when the lead was removed by chelation at this stage, renal clearance returned to normal with no apparent evidence of permanent damage. From this it would appear that in young animals the little storage of lead that occurs at low level intake may not be of clinical significance and that, providing there is no continuing high-level exposure, on current evidence there would appear to be little risk of lead intoxication or possible permanent damage in a normal healthy child whose fundamental nutrition is in all respects adequate.

With reference to Dr McCallum's comment on impotence, I recall a case in which a man of magnificent physique was referred by his general practitioner to a hospital, where the complaint of impotence was about to be diagnosed as psychological, when somebody suggested that his lead excretion had not been tested. On examination this was found to be extremely high and at this stage the case was referred to me as he worked in a scrap metal recovery plant in my area. The housekeeping of the works was appalling and with the aid of the Alkali and Factory Inspectorate we were able to close the works. During his employment there is little doubt that not only he but all others who worked there must have been exposed to virtually uncontrolled lead fume and lead-containing dust, and it would have been interesting, had the present climate of opinion existed at the time, to do a full survey of these workpeople and their families.

Dr P. S. I. Barry: During 1966–69 we studied blood lead and urinary lead concentrations in occupationally exposed and unexposed pre-employment subjects, comprising a total of 911 male adults, in a lead alkyl industry (Figs C and D).

Ninety-nine per cent of the unexposed group had blood lead concentrations below 60 $\mu g/$ 100 g of blood and 99 per cent of the exposed group had blood lead concentrations below 80 $\mu g/$100 g of blood, whereas the distribution of lead in urine concentrations showed 99 per cent of the unexposed group with urinary lead concentrations below 90 $\mu g/l$, and 99 per cent of the exposed group below 170 $\mu g/l$.

A crude correlation exists between lead concentrations in blood and urine. In general, the higher the blood lead concentrations the higher is the urinary excretion rate, but for a given blood lead level the urinary excretionary rate of lead is higher for occupationally exposed subjects than for those not occupationally exposed (Fig. E).

The mean values for our non-occupational group are 29·6 μg Pb/l of urine and 22 μg Pb/ 100 g of blood compared with 65 μg Pb/l of urine and 41 μg Pb/100 g of blood in our occupationally exposed group. The overall mean level of atmospheric exposure of the occupational group was of the order of 100 μg Pb/cu. m air. Over a period of some 20 years, no ill-health attributable to this level of exposure has been observed.

Concentrations of lead in blood and urine remained substantially constant with age and also with length of employment in the industry.

In conclusion, our study has demonstrated a two-fold increase of lead in blood and in urine of a population group exposed to lead where the atmospheric concentrations exceeded the concentrations of lead in city air by more than 20-fold. We hope that these findings may lend perspective in relation to the contribution of lead in the ambient air to the total body burden of lead.

Professor T. P. Whitehead: It is obvious that we have some captive populations of people with obvious lead intoxication, and we have had them for many years. In order to detect whether there are symptoms and conditions which we are missing and vague psychiatric and depressive states that may conform with lead ingestion, industry as a whole is making very careful studies on this. I know they study the obvious cases of lead poisoning but are there very complete histories available of these workers?

Dr P. S. I. Barry: I would say that in the major lead industries the answer is "yes". You probably could not say this about them all, but certainly it is so in the larger lead industries, which are subject in any case to legislation in respect to medical examinations. In our own industry, we are not subject to legislation in this sense. We undertake examinations every two months for our employees and give each employee a pre-employment medical examination which is pretty thorough. We are able to watch, at fairly frequent intervals,

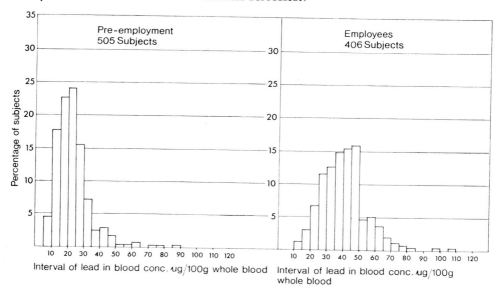

FIG C. Distributions of lead in blood concentrations.

the men as they go through their lives with the company. It is perfectly true that from time to time we do see people suffering from depression. We see people suffering from all sorts of other conditions but in the assessment we attempted to make some years ago with respect to the depressives we did not find any greater morbidity in our group compared with that of the general population.

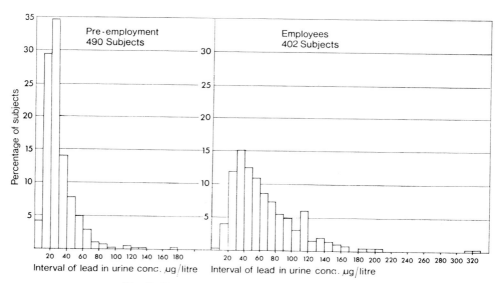

FIG D. Distributions of lead in urine concentrations.

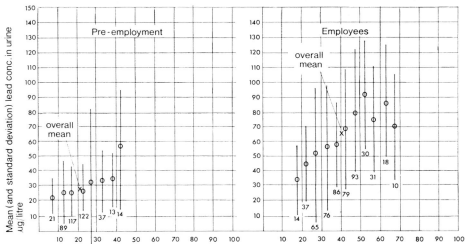

FIG E. Correlations of lead in blood and lead in urine.

Professor D. Bryce-Smith: Sir, could I just butt in briefly? I submit that Dr Barry's figure of 20-fold exposure should be divided by three because his subjects are exposed only for 40 hours in a week.

Dr P. S. I. Barry: I made a calculation on this because I had a feeling it might come up. If you take 20 cu. m as the amount of air breathed per day and its lead content as 2 μg/cu. m—the figures given by Dr Barltrop—the average non-occupationally exposed individual breathes in 280 μg Pb/week. Lead content of the air in city streets may be 5 μg/cu. m but I submit that people do not sleep on city streets. Making a further assumption that exposed workers breathe 10 cu. m of air while working and 10 cu. m while resting, the lead breathed in while exposed for five working days is 5000 μg, on top of which there is 180 μg while not working, giving a total of 5180 μg. This is a ratio of 1 : 18·5.

Dr R. Stephens: In response to Dr Barry's comments, I think it is important to remember that the airborne lead in a lead smelting area is not produced by an automobile engine. Particle size distribution is vital in this issue. With respect to Mr Greaves, he refers to airborne lead solely, and compares this with total ingested lead on an average. I submit that you can only use this argument if you know the ultimate fate of all the airborne lead derived from petrol lead.

Dr R. J. Sherwood (Esso Europe): I wonder if Professor Bryce-Smith would mind quoting the full context with reference to the American publication "Airborne Lead in Perspective". He quoted a section out of it. He did not quote the preamble, nor at the end, where they said that, on the basis of available epidemiologic evidence, it is not possible to attribute any increase in blood lead concentration to exposure to ambient air below a mean lead concentration of 2 or 3 μg/cu.m. and that only special small groups of people in large cities are exposed to higher mean atmospheric concentrations.

The full text of this section reads as follows:

"Only in the urban setting is man possibly exposed to hazardous circumstances relative to lead pollution, occupational exposures in the lead-using industries excepted. The exposures are mainly the consequences of atmospheric emissions. The high concentrations of lead in urban air and on the surfaces of parks and streets constitute a source of intake additional to the usual dietary sources and in special circumstances may be a substantial source. The contribution of atmospheric sources of lead to the total body burden of city dwellers varies considerably, and depends on the particular urban complex, and on the place of residence and of work within that complex. An extensive survey of men in three large urban complexes has suggested, but not shown with definitive evidence, a strong association between the concentration of lead in the blood and time spent in areas of high automobile traffic density. Although this evidence of the impact of lead inhalation is inferential, limited experimental data are consistent with the conclusion that the amount of inhaled lead is about one-half (and in special circumstances twice) the amount that comes from diet, depending on the particular urban micro-climates encountered in the course of daily activities. However, it is not possible, on the basis of available epidemiologic evidence, to attribute any increase in blood lead concentration to exposure to ambient air below a mean lead concentration of about 2 or 3 μg/cu. m; only special small groups of people in large cities are exposed to higher mean atmospheric concentrations."

Dr P. S. I. Barry: Could I ask the experts present whether significant changes occur in the ALA-D activity at blood lead levels below, say, 20 μg/100 ml?

Dr A. E. Martin: At these low levels I do not think we have anything with which to compare it. The level of enzyme activity declines until one gets to the lowest level, so that we are comparing the level of very low enzyme activity at the very low blood levels with the higher ranges of normality and the ranges which are above normality.

Pointers to Future Studies

Professor T. P. Whitehead: Our time is up. I wonder if one or two people would like to make contributions in a very few words about pointers to where research should be going. We have heard Dr Martin and Dr Glover speak about one or two projects that are going forward. We have the possibility of the study of lead toxicity in industry. I wonder if we could have other proposals?

Dr R. F. Crampton (BIBRA, Carshalton): I would like to put in a little plea for sanity as far as research for the future goes. Everyone knows that lead under certain circumstances is dangerous and poisonous. The Commissions are aware of the neurocrosis of lead. We had a reference this morning to the possible effects on testicular development. Incidentally, if the very low quantities of lead did produce testicular atrophy, this would hardly be accompanied by a teeming horde of aggressive adults, fully exposed to lead, but my plea for the future is, really, that we do know certain sections of the population have been exposed to high levels of lead. We may identify "hot spots" of children and other sections of the population who are, have been, or will be exposed. Sooner or later, someone will bring up the problem of long-term effects rather than acute toxicity, and if these populations could in fact be kept in mind and monitored long-term, I do not mean weekly or monthly, I mean yearly over a period of ten or 20 years, and it would take very little effort to do, the results might, in fact, be very reassuring, or otherwise, in contributing to knowledge in the long term.

Dr J. D. Butler (University of Aston in Birmingham): I think we are all aware that very shortly we are going to have the link-up of a motorway system, the M5, the M6, and the M1. This is a region where we are going to have a great deal of vehicular traffic. The City of Birmingham, particularly the Medical Officer of Health, who is here today, is particularly conscious of this, and the future programme, which has already in fact started, for the monitoring of lead and other motor vehicle emissions is in progress at the present time and we hope that if funds can be found to sustain this programme that it will continue for at least two years and that it will contribute something of an environmental help to us in order to attempt to find out just exactly what the hazards are associated with the motorway programme.

Conclusion

Professor T. P. Whitehead: It would be impossible for me to attempt to summarize the whole proceedings but I feel that I, as a non-expert, have sensed some developments.

—It is obvious that the oil and motor industries are making a real effort in research to combat the problem of lead emissions. I have heard that the Germans are going unilateral on lead control and Dr Stephens kindly sent me a report about New York proposing to ban lead in petrol early in 1974. Whether the effort is directed towards new engines and lead-free fuels, or towards continued use of lead and preventing it from reaching the atmosphere, is something that industry is best left to decide.

—I have been convinced that there are people in the U.K. who are in an environment where the combination of lead from a variety of sources is conducive to sufficient lead being ingested to give them clinical symptoms. Apart from lead in air, we have heard that lead may appear in water. It is certainly in soil and in the pica activities of children. This is already being investigated but the effort needs to be stepped up, particularly in areas of heavy traffic density and petrol engine activity.

—I think that lead is going to be with us for a long time. There must be a lot more investigation of its toxicity, the relationship between enzymes and blood lead levels, and what they mean in terms of health.

—There needs to be more of the type of work that Professor Lawther is doing. That is to say, finding out how lead gets into the atmosphere, whence we breathe it, and how it gets into the human organism. More effort in this direction should pay off in other areas of pollution because the systems and methods will be applicable.

—I feel we cannot wait years for epidemiological studies but that more effort should be devoted to the epidemiology of those who are known to be exposed to lead.

I would like to conclude by thanking all of you who have taken part in a very interesting meeting. This conference was organized under the auspices of the three societies, who provided the chairmen for the three sessions but the Administrative Secretary and staff of the Institute of Petroleum handled all the arrangements and I think they have done it extremely well. I thank them on your behalf.

We have had expression of a variety of opinions, we have done it with good manners, and have respected each other's opinions. I hope that it will be a pointer to other conferences of this type on pollution because it has not been just one group of people shouting that something should be done. We are all concerned in this matter and I am certain that the way to handle it is in getting together as we have done today and understanding each other's point of view. I hope that we will give a lead to other countries to do it this way.

REFERENCES

1. Gordon, I., and Whitehead, T. P. *Lancet*, 1949, **2**, 647–50.
2. Whitehead, T. P., and Prior, A. P. *Lancet*, 1960, **2**, 1343–44.
3. Transactions of the 33rd Annual Meeting of the American Conference of Governmental Industrial Hygienists, Toronto, Canada, 24–28 May 1971.
4. Williams, M. K. King, and Walford J. *Brit. J. Industr. Med.*, 1969, **26**, 202.
5. Factories Act 1961 (c. 34) s. 75 (1) (b) (e).
6. Fussell, D. R. *Petrol. Rev.* 1970, 192–202.
7. U.S. National Academy of Sciences Report "Airborne Lead in Perspective", 1971.
8. Final Report of Departmental Committee on Ethyl Petrol. Ministry of Health, London, 1930.
10. Addicott, J., *et al.* I. Mech. E. Paper C 133/71.
11. Galzigna, —, *et al. Clinica Chimica Acta*, **26** (3), 391–3.
12. B.P. 1,252,228.
13. Hudnall, J. R., *et al.* "New Gasoline Formulations Provide Protection Against Corrosive Engine Wear". SAE Paper, 690,514, Chicago, 19–23 May 1969.
14. Leake, J. P., *et al.* "Use of Tetra-ethyl Lead Gasoline in its Relation to the Public Health". U.S. Public Health Bulletin No. 163, 1926.
15. Cremer, J. E. *Brit. J. Industr. Med.*, 1959, **16**, 191–99.
16. Cremer, J. E. *Brit. J. Industr. Med.*, 1961, **18**, 277–82.
17. Cremer, J. E. *Ann. Occup. Hyg.*, 1961, **3**, 226–30.

18. Springman, E., Bingham, E., and Stemmer, K. L. *Arch. Environ. Hlth*, 1963, **6**, 469–72.
19. *Paediatrics*, August 1969, **44** (2), 291.
20. Perlstein, M. A., and Attala, R. *Clin. Paediat.*, 1966, **5**, 292.
21. Stopps, —. *J. Washington Acad. Sci.* 1971, **61**, 103.
22. Goodman, G. T., and Roberts, T. M. *Nature, Lond.*, 1971, **231**, 287–92.

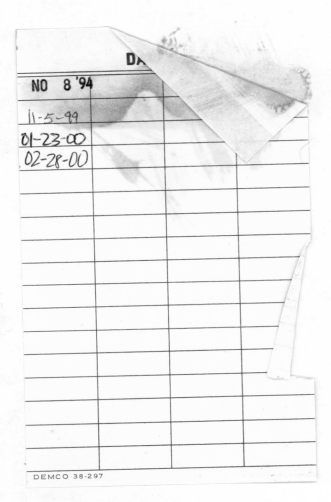

DA

NO 8 '94

11-5-99

01-23-00

02-28-00

DEMCO 38-297